Causes and Consequences of Map Generalisation

To Andrew and Ana

Causes and Consequences of Map Generalisation

ELSA MARIA JOÃO

Department of Geography and Environment,
London School of Economics

TAYLOR & FRANCIS
ALERE FLAMMAM
1798 – 1998

| UK | Taylor & Francis Ltd, 1 Gunpowder Square, London, EC4A 3DE |
| USA | Taylor & Francis Inc., 1900 Frost Road, Suite 101, Bristol, PA 19007 |

British Library Cataloguing-in-Publication Data

A catalogue record for this book is available from the British Library.

ISBN 0-7484-0777-4 (cased)
ISBN 0-7484-0776-6 (paperback)

Library of Congress Cataloging-Publication-Data are available

Cover design by Hybert Design and Type, Waltham St Lawrence, Berkshire
Typeset in Times 10/12 pt by Santype International Ltd, Salisbury, Wiltshire
Printed in Great Britain by T. J. International Ltd, Padstow

Contents

Series introduction

▊

Research Monographs in Geographical Information Systems

Welcome

Research Monographs in Geographical Information Systems aim to provide a publication outlet of the highest quality for research in GIS which is longer than would normally be acceptable for publication in a journal. The series will include single- and multiple-author research monographs, possibly based upon PhD theses and the like, and special collections of thematic papers.

The Need

We believe that there is a need, from the point of view of both readers (reseachers and practitioners) and authors, for longer treatments of subjects related to GIS than are widely available currently. We feel that much research is actually devalued by being broken up into separate articles for publication in journals. At the same time, we realise that many career decisions are based on publication records, and that peer review plays an important part in that process. Therefore, a named editorial board has been appointed to support the series, and advice will be sought from them on all submissions.

Successful submissions will focus on a single theme of interest to the GIS community, and treat it in depth giving full proofs, methodological procedures or code where appropriate to help the reader to appreciate the utility of the work in the monograph. No area of interest in GIS is excluded, although material should demonstrably advance thinking and understanding in spatial information science. Theoretical, technical and application-oriented approaches are all welcomed.

The medium

In the first instance, the majority of monographs will be in the form of a traditional book, but, in a changing world of publishing, we actively encourage publication on CD-ROM, the placing of supporting material on web sites, or publication of programs and of data. No form of dissemination is discounted, and prospective authors are invited to suggest whatever primary form of publication and support material they think is appropriate.

The editorial board

The monograph series is supported by an editorial board. Every monograph proposal is sent to all members of the board, which includes Ralf Bill, António Câmara, Joseph Ferreira, Pip Forer, Andrew Frank, Gail Kucera, Enrico Puppo and Peter van Oosterom. These people have been invited for their experience in the field, of monograph writing, and their geographical and subject diversity. With particular monographs, members may also be involved later in the process.

Future submissions

Anyone who is interested in preparing a research monograph should contact either of the editors. Advice on how to proceed will be available from them, and is treated on a case-by-case basis.

For now, we hope that you find this, the second in the series, a worthwhile addition to your GIS bookshelf, and that you may be inspired to submit one too.

Editors:	Dr Peter Fisher	Dr Jonathan Raper
Addresses:	Department of Geography	Department of Geography
	University of Leicester	Birkbeck College
	Leicester	7–15 Gresse Street
	LE1 7RH	London, W1P 1PA
	UK	UK
Telephone:	(+44)(0) 116 252 3839	(+44)(0) 171 631 6457
Fax:	(+44)(0) 116 252 3854	(+44)(0) 171 631 6498
E-mail:	pffl@le.ac.uk	j.raper@geog.bbk.ac.uk

Foreword

It is a great pleasure to welcome the publication of this monograph. There are two main reasons for this – the first largely personal, and the other much more a professional one. As to the first, it is always a pleasure when a former PhD student achieves recognition in such a highly visible way. In the early 1990s, Elsa João and I worked together on a project on generalisation funded jointly by the Economic and Social Research Council and the Natural Environment Reseach Council; Professor Stan Openshaw and Dr Brian Kelk were also much involved, as was Graeme Herbert. From part of that project came Elsa's PhD thesis and this monograph. My departure from academia to Ordnance Survey resulted in Dr Jonathan Raper taking over the supervision of Elsa's PhD thesis – with manifestly successful results.

More profoundly, Elsa's work begins to fill in one of the great lacunae of generalisation, perhaps the greatest remaining research problem in Geographical Information Systems and computer-based cartography. Traditional cartographers have long practised elegant manual generalisation of maps. More recently, non-traditional cartographers have written learned treatises about the theoretical or algorithmic bases of such generalisation. But remarkably little has been known in quantitative terms about the magnitude of past generalisations. This is of fundamental importance, since virtually all GIS make use of standard 'framework data' produced by national mapping organisations, which contain substantial manual generalisations. Most other data collectors assemble their own data on one map scale or other of these national frameworks. Originally prepared for paper-based mapping, the generalisation effects have become embalmed in the digitally encoded versions of the original maps.

Only measurement of the magnitude of these effects can tell us whether such generalisation is likely to have significant 'downstream' effects when data drawn from different map sources are combined and added value – the 'something for nothing' promise of GIS – is extracted. Given this, it will be obvious why I welcome the publication of this empirical research.

DAVID RHIND
Director General
Ordnance Survey

Preface

This book is the culmination of a four-year research project that I carried out at Birkbeck College, London, on map generalisation. The divergence between maps for the same region at different scales, such as changes in the number of features represented and changes in measured lengths, is known as generalisation effects. The subject has become a major issue among GIS researchers, and this book aims to provide for the first time a detailed *quantitative* analysis on the consequences of map generalisation.

Although this study has some relevance to conventional cartographic techniques, it is ultimately of greater pertinence to the GIS community, where the ability to quantify generalisation could be made easier by automation and where the consequences of uncontrolled generalisation are more serious. I propose (as others have) that automatic generalisation methods should be used to a greater extent, for both analysis and display purposes, in order to control generalisation effects. Automated generalisation should enable the quality of generalised cartographic products to be more closely controlled. This should entail a quantification and a storage of generalisation effects that can then be carried forward and taken into account in subsequent GIS map manipulations.

The problem of generalisation has dogged cartographers for many years and is likely to do so for many more. By writing this book, I intend to throw light on the very real effects of generalisation on maps and on the use of GIS. Wherever possible, I have tried to keep technical terms to the minimum. My hope is for readers of this book never again to take for granted a particular scale of a map.

The book is organised into seven chapters. The problem of map generalisation and scale, both from a traditional cartographic perspective and a computerised approach, is introduced in chapter 1. The previous work done on the study of generalisation effects (on paper and digital maps), and on the consequences of generalisation within GIS is reviewed in chapter 2. The maps used in this study and all of the data treatments carried out are described in chapter 3. In chapter 4 it is explained how generalisation effects were separated out from other sources of map error, and the method developed to measure generalisation effects locked into maps is described. The results of the quantification of generalisation effects on manual and automated generalised features are presented in chapter 5, and in chapter 6 I evaluate how these effects have impinged on typical GIS map manipulations. Finally, in chapter 7, the results obtained in this study are discussed, the methodologies used for measuring the generalisation effects are evaluated, and ways to minimise and keep track of generalisation effects within a GIS are proposed.

I would like to take the opportunity to thank all of the individuals and organisations that made this research possible. I start by expressing my thanks to Professor David Rhind (now Director General of the Ordnance Survey) and Dr Jonathan Raper for their help, constructive criticism and advice. Special thanks must also go to my husband, Andrew Thompson, who was a constant source of analytical criticism and support, from which this book has benefited. David Livingstone wrote the compiler program that determined the fractal dimension of lines, and therefore his help was particularly important and appreciated. Discussions with Graeme Herbert, who shared part of the generalisation research with me for two years, led towards a better understanding of generalisation effects.

This study was in part carried out under one of the research projects of the ESRC/NERC (the United Kingdom Economic and Social, and the Natural Environment Research Councils) joint programme on Geographical Information Handling. ESRC/NERC funded part of this research under grant GST/ 02/490. In parallel with this work, a group at the University of Newcastle upon Tyne were working on the development of a new line generalisation algorithm and on modelling error propagation in GIS. The interchange of ideas with this group was very fruitful, and I take this opportunity to express my thanks to the group members, especially Professor Stan Openshaw.

The Ordnance Survey kindly provided the data for this research, at no cost, and the Instituto Português de Cartografia e Cadastro permitted use of their maps and provided information on how they were compiled. Many staff from these two institutions patiently answered my queries: I thankfully acknowledge their help. Laser-Scan Ltd, of Cambridge, gave considerable assistance to the work, notably in the scanning of the Portuguese maps and in the solving of technical problems. I would also like to thank the many generalisation researchers around the world, too numerous to mention individually, for their comments and advice. Finally, I gratefully acknowledge the staff of

Birkbeck College for all their support and for the full use of the facilities of the Department of Geography.

ELSA JOÃO
London

CHAPTER ONE

Scale and generalisation in GIS

Generalisation is an inherent characteristic of all geographical data. All maps, whether digital or analogue, are generalised representations of reality; and the more generalised a map is, the more remote from reality it becomes. Generalisation denotes a process by which the 'presence of phenomena or events in a referent space are essentially reduced and/or modified in terms of their size, shape and numbers within map space' (Balodis, 1988, p. 71). The end product of this generalisation process is a derived data set with less complex and usually more desirable properties than those of the original data set.

However, the generalisation process results in a dilemma. On the one hand, it is necessary to generalise in order: to improve the display quality of a map at a scale smaller than the one from which it was compiled; to allow analysis with different degrees of detail; and to reduce data storage requirements. On the other hand, generalisation also causes unintended transformations of the data (such as changes in measured lengths and areas) that can alter the topology of geographical phenomena and affect subsequent statistical or geometrical calculations. Normally, the user would wish to minimise the unforeseen and unquantified effects of generalisation. It is the study of these generalisation effects that is the subject of this book.

The manner in which generalisation affects the statistical and geometric properties of spatial data is fundamental to the use of Geographical Information Systems (GIS). These are computer-based systems that are increasingly being used to store and manipulate geographical information. The main advantage of GIS over paper maps is their ability to carry out sophisticated and extensive analysis, and their success in doing so depends on keeping control of unintended data transformations. If the results from combining two or more data sets are spurious because of the poor quality of the data – rather than of the situation in the real world that the data are supposed to represent – then all subsequent interpretations and actions based upon them are liable to be flawed. The oft-quoted advantage of GIS as a means of data integration becomes problematic when data compiled from different sources or scales fail to match in a logical fashion (Rhind and Clark, 1988), leading, for instance, to the improbable situation of soils supposedly located in the sea, or river channels located along the side of a ravine. Unfortunately, most generalisation errors are more difficult to spot.

The research in this book measures the magnitude of generalisation effects already 'locked into' paper maps and compares these with transformations by GIS-based generalisation algorithms. By carrying out analyses using different source scale data, it is possible to quantify just how important these generalisation effects can be. On the basis of this information, ways of improving the generalisation process within a GIS are proposed. Although this work has relevance to conventional cartographic techniques, it is expected to be more pertinent to GIS and geographical database use. The ultimate objective of this work is to provide the key by which future GIS users will be able to generalise geographical information according to their own need and criteria of quality – but to carry this out via an automated procedure, so that the user's possibly limited skills are not a constraint – and to *quantify* the effects of generalisation on the data and subsequent analyses. Finally, in this book attention is drawn to the best ways of applying generalisation within a GIS, by controlling the unintended consequences of generalisation.

A multitude of factors are involved when generalising from one spatial data set to another data set with less detail:

- The purpose for which the map is going to be used.
- The geographical region that needs to be mapped.
- The original and the final map scale.
- The particular individual undertaking the generalisation.
- The taste and knowledge of the user or client.

All of these elements will shape the form of the final product, and will determine the type and the amount of generalisation effects 'locked into' the maps. One of the most important factors though, in determining which features are maintained and how much they are modified, is the scale transformation.

1.1   Scale and resolution

Almost always, maps take up less physical space than the area that they represent. This reduction in size is reflected by the scale of the map, and is usually represented as a ratio or a graphic scale. The ratio indicates to the map user the number of metric units on the ground that are represented by a unit in the space of the map model. Scale is considered to be the single most crucial mathematical feature of any map and, because of its importance, scale is often used as a primary means of categorising maps – from large-scale maps of 1 : 10 000 to atlases of 1 : 1 million.

Often associated with scale is the term *resolution*: 'The resolution of a data set defines the smallest object or feature which is included or is discernible in the data' (Goodchild, 1991, p. 113). Although scale and resolution have distinctive meanings, they are closely linked because for each map scale there is a lower limit to the size of an object that can be usefully shown on a map. For topographic maps and dark printing colours, the minimum size of a point that is still discernible to the human eye has a diameter of 0.25 mm (Swiss Society of Cartography, 1987). On the basis of this minimum resolution, the minimum size of the objects that can be shown true to scale on a map will vary for different scales. For example, at the scale 1 : 10 000 the smallest object would be 2.5 m long, while at the scale of 1 : 250 000 the minimum size increases to 62.5 m.

'It is impractical or functionally impossible to collect data using a one-to-one correspondence between cartographic entities and objects: resulting maps would be a replication of the real world, not a model of it' (Cromley, 1992a, p. 132). The choice of the scale at which data should be collected needs to be determined at the outset as it must reflect the objectives of the map-maker, the precision of the instruments used in the original survey, and the type of spatial processes being modelled and analysed. The resolution of the instruments used sets the limit, because it is not possible to measure features more precisely than the equipment allows.

When the scale of a map is decreased, there is less physical space in which to represent the geographical features of a region. As the process continues, the features will need to be exaggerated in size in order to be distinguishable at a smaller scale. As geographical features 'fight' for representation in the reduced map space, some features will need to be eliminated, and those remaining may be further simplified, smoothed, displaced, aggregated or enhanced. In the extreme case, the map loses its geometric properties and becomes a caricature. Choice of scale, therefore, 'sets a limit on the information that can be included in the map and on the degree of reality with which it can be delineated' (Robinson and Sale, 1969, p. 15). Thus objects such as houses could be reproduced readily at 1 : 1250, 1 : 2500 or even 1 : 10 000 scale, but the constraints of human drawing and perception ensure that a number of the houses have to be grouped together into blocks, or eventually urban areas, for depiction at smaller scales.

With paper maps, different scales have always been represented by physically distinct maps. However, with the advent of GIS, to an extent, the representations have become independent of scale (Frank, 1991). With GIS it is possible to 'zoom in and out' of the map, producing a continuous series of scales as desired. However, the scale at which the original data were *digitised* is still a limiting factor for the amount of features, shapes, level of detail, and so on, that the user can see. When the user 'zooms in' beyond this scale, more detail does not miraculously appear; eventually, all that remains is a blank screen. Conversely, 'zooming out' from a map can result in too much detail being presented, and finally the collapse of all features into a single point. More importantly, the generalisation effects contained within the manually generalised paper map are 'locked into' the digital versions of the maps. As digital maps are usually converted from traditional paper maps, at the very least, their intrinsic generalisation is determined by the manual cartographers who prepared the original versions of those maps. Therefore, the *source scale* of any digital map remains a crucial defining characteristic and one which will ultimately affect the map's accuracy, precision and quality control.

1.2   The cartographer, the map and the region

Most generalisation is carried out manually by trained cartographers, their work and research being driven primarily by the need to show features at a map scale much smaller than that at which the information was originally presented. The cartographers' experience, natural intuition, possible knowledge of the area and taste all influence how they generalise a map; in other words, their selection of how to draw what, with what priority and detail, and which attributes to display as text. Manual generalisation is accomplished using basic cartographic rules and techniques, and the cartographer's ability to *view* the map as a whole and for what it represents.

Because individual cartographers use their *own* knowledge and judgement to carry out generalisation, the practice of manual generalisation is often described as *subjective* (Keates, 1989). The subjectivity of the process is exacerbated by a further aspect that cartographers have frequently evoked – the need to produce an end result that has some aesthetic qualities (Robinson, 1989). Consider, for instance, the views of Eduard Imhof (1982, p. 86), former President of the International Cartographic Association, on producing the perfect map:

> . . . the greatest possible accuracy, with respect to the scale of the map; clear expression of metric information; good characterisation in the forms; the most naturalistic forms and colours; the greatest possible clarity of meaning and good legibility, simplicity and clarity of graphic expression; and finally, summarising all these qualities, a beauty peculiar to the map itself.

Generalisation as art has been proposed since the beginning of the twentieth century. According to Eckert (1908, pp. 346–7), 'As long as the scale allows the objects in nature to be represented in their true proportion on the map, technical skill alone is necessary. Where this possibility ends, the art of the cartographer begins. With generalisation, art enters into the making of maps'.

The way in which cartographers work will reflect the cartographic traditions of their country or the specific rules of the mapping agency. The map purpose will also affect the underlying classification of which features are the more important. Obviously, a road map will give predominance to roads and will be less concerned with the way in which rivers are portrayed. The magnitude and type of the generalisation process also changes with the geography of the region. The concentration of features and the type of spatial data influence how generalisation affects each feature. The cartographer, for example, is likely to want to retain an isolated building in the countryside, while the same sized building might be removed if located in a town. A region with a large concentration of rivers might proportionally have more rivers eliminated than a region with only a few rivers present. Also, the more densely that features are concentrated together, the more they need to be generalised. The classical example of this is the case in which a river, a railway and a road pass through a narrow valley. In this case, one or more of these features needs to be displaced at the smaller scale, in order that all features can be displayed without overlapping each other. This displacement would not be necessary if all of the features were situated far apart.

Map generalisation as done by manual cartographers requires adaptability to all of these different circumstances, from the characteristics of the particular geographical region that needs to be mapped to the differing purposes for which the maps are going to be used. Because of the multitude of factors that need to be taken into account in order to generalise, it would seem that a computer-based system (such as a GIS) could *potentially* offer a good solution to the problem. A computer-based system would also have the added benefit of an iterative capacity — the ability to redo a process until an acceptable solution is found. However, although there is potential for GIS to improve certain generalisation tasks, the problem has proved far too complex to date for a system to have been developed that can successfully handle generalisation in its entirety.

1.3 Generalisation and the computer

Over the past 25 years, there has been a drive towards automating map generalisation. On the whole, traditional cartographers have always coped with generalisation, but as GIS have become more prevalent, so the issue of generalisation has increased in importance: 'Generalisation is perhaps the most intellectually challenging task for the cartographer, a proposition supported by the comparatively marginal success of computer algorithms in generalising maps' (Monmonier, 1982, p. 170). Fifteen years on, automation of the handling of map generalisation procedures remains one of the major challenges in GIS research. Due to its complex, diverse and non-deterministic nature, the generalisation process has proved to be very difficult to automate, particularly because one is attempting to mimic such a subjective and intuitive procedure. This explains why, to date, all GIS fail to take adequate account of the problem of generalisation.

In a GIS environment there is more than one reason why users would wish to generalise their data. Generalisation for display purposes is carried out with the primary goal of communicating visual information clearly. Unwin (1981), for example, points out that if the same contour interval is kept with decreasing map scale (e.g. 10 m), the slopes will have the *appearance* of being steeper. The use of generalisation to improve the display properties of a map is an objective shared both by the GIS community and the traditional cartographers. However, in the case of automated generalisation, the reduction of scale is not the only motivation for generalisation: 'In the context of digital cartographic systems and GIS, generalisation has obtained an even wider meaning' (Weibel, 1995a, p. 259). Two extra requirements for GIS generalisation are especially pertinent: to reduce data storage requirements and to manipulate data for analysis.

A GIS user can decide to generalise data in order to reduce the amount of data storage space and processing time, driven either by financial or technological constraints. More importantly, generalisation within a GIS is also performed for analytical purposes: 'The necessity to understand at which scales or range of scales spatial processes occur is one of the driving forces behind generalisation today' (Müller, 1991b, p. 457). It does not necessarily follow that the use of a larger scale leads to a more appropriate analysis. Sometimes, features or patterns emerge only *after* generalisation, and only then can they be analysed. This concept of scale-dependent phenomena is encapsulated, for example, in the phrase 'when you can't see the wood for the trees'. Therefore, as suggested by Moellering and Tobler (1972), in some circumstances a higher level of generalisation may be able to provide more explanation about the variance of a spatial variable than finer resolution levels. It is especially important that, for analysis purposes, generalisation procedures preserve data characteristics as much as possible (this is covered in more detail in section 2.2).

There are several reasons why the GIS community seeks to automate generalisation. Not all users are skilled in map generalisation procedures, and the number of GIS workers (and hence map-makers) without cartographic training is growing rapidly. Also, many GIS users are unaware of the effects of generalisation carried out within 'black box' GIS or already enshrined in data sets. As a consequence, many users waste time 're-discovering' generalisation effects. Finally, and most importantly, if it was possible to generalise automatically, then GIS could store just *one* single large-scale data set which could be generalised whenever needed. Update procedures would be easier to carry out on a single data set rather than on the multiple source-scale data sets which are currently stored in most GIS. (Multiple representations versus scale-independent GIS are discussed in more detail at the end of this chapter.)

As with manual generalisation, automated generalisation is taken as a process of selecting and simplifying a description of geographic phenomena. However, unlike manual generalisation methods (which involve the simultaneous application of different factors by the cartographers), in order to automate generalisation it is necessary to break the process up into a series of smaller, identifiable steps and, eventually, into algorithms (McMaster, 1987). Each one of these steps is termed a 'generalisation operator', which defines the transformation that needs to be achieved (e.g. line simplification). For each one of these operators, generalisation algorithms are used to implement the particular transformation. An algorithm is a set of rules that specify a sequence of actions to be taken to solve a problem. Most importantly, each rule should be defined so precisely and unambiguously that in principle it could be carried out by a computer (Walker, 1988). It is the identification of these rules and 'their implementation into a system, which can simulate the work of a traditional cartographer, that is one of the most difficult challenges facing the GIS research agenda of the 1990s' (Müller, 1991b, p. 457).

A classification of generalisation operators is shown in Table 1.1. The table is divided into raster- and vector-mode generalisation, on the basis of the distinction between raster (where the location of geographical features in a regular grid is defined by the row and column that they occupy) and vector (where points and lines mark out the position of the features, with the position of the points stored as unique coordinate values) GIS. This is not a definitive list, but it is an illustration of the types of generalisation operators that have been described in the literature. Currently, there is still no agreement on a comprehensive set of generalisation operators. In fact, some confusion remains on what the different operators mean. A study by Ricger and Coulson (1993) found that not all authors defined their generalisation operators and, where definitions were given, major differences existed, with authors using different terms for the same definition and using different definitions for the same term. In order to redress this problem, the Working Group on Generalisation of the OEEPE (Organisation Européenne d'Etude Photogrammétriques Experimentelles) is coordinating an international research effort that will eventually lead to the existence of a common terminology for generalisation operators (see Ruas, 1995b, for a draft version).

In the case of spatial data, generalisation in a vector mode can be subdivided according to different types of data. Unwin (1981) describes four types of geographical features within a vector format: points, lines, areas and volumes (i.e. surfaces). In addition to these four types, João et al. (1990) describe a further category. This is flow data, which indicate movements between places (e.g. migration of population) and is probably best conceived geometrically as multiple sets of individual origins and connected destinations. All five of these geographical data types require different generalisation operators, although the different categories are not necessarily rigid. An area can be generalised into a point or a line, and a group of points can be converted into an area (see Monmonier, 1991).

Spatial data are only one aspect of vector-mode generalisation. There are also attribute data, which can be either thematic or temporal (see Table 1.1). Generalisation of *thematic* data, such as the classification of feature types

Table 1.1 A classification of generalisation according to raster- and vector-mode operators.

Raster-mode generalisation (McMaster and Monmonier, 1989)	Vector-mode generalisation	
	Attribute data (Brassel and Weibel, 1988; McMaster, 1989)	Spatial data (McMaster, 1989; Monmonier, 1991; and inferred from Tobler, 1989)
Structural generalisation	Thematic data	Punctiform data
simple structural reduction	classification	select/omit
resampling	symbolisation	aggregate
tesselation change		displace
vector-to-raster conversion	Temporal data	
Numerical generalisation		Linear data
low-pass filters		enhance
high-pass filters		simplify
compass gradient masks		smooth
band-rationing		displace
vegetation indices		merge
		select/omit
Numerical categorisation		
minimum-distance-to-means classifier		Areal data
parallelepiped classifier		collapse
maximum-likelihood classifier		simplify
		select/omit
Categorical generalisation		smooth
merging (of categories)		amalgamate
aggregation (of cells)		displace
non-weighted		segment
category-weighted		enhance
neighbourhood-weighted		
attribute change		Volumetric (surface) data
feature dissolve		smooth (trend surface)
minimum-size suppression		enhance
categorical suppression		simplify
erode smoothing		
gap bridging		Flow data
		amalgamate
		select/omit

(e.g. deciduous and coniferous into woodland), can be automated in a straight-forward fashion, but generalisation in the *temporal* domain has received only minor attention in cartography. An example of temporal generalisation would be a coastline slowly eroding over time. Is its position defined on a map every week, month, year or decade? According to Langran (1992), temporal scale change is likely to be simpler than spatial scale change because of time's single dimension. Temporal data are comprised of points (the events) and lines which connect them (the states), and therefore in certain circumstances these points could be suited to a line generalisation approach (such as the Douglas–Peucker algorithm) which selects values that deviate from a norm by some established tolerance (Langran, 1992). The author suggests this would be appropriate when the points stored by the system are samples of a continuous process – for example, the eroding coastline – but are surveyed only period-ically. Langran even claims that 'generalising or interpolating over time could easily provoke a level of research excitement comparable to that generated by spatial generalisation and interpolation' (Langran, 1992, p. 36).

A key aspect of Table 1.1 is the distinction between raster and vector oper-ators. However, these two operators are not necessarily mutually exclusive, as both types could coexist in a given GIS. Monmonier (1986) suggests the design of an operational system for automated line generalisation based on a *hybrid* data structure; that is, combining vector- and raster-type operators. According to Monmonier, generalisation of linear data (as opposed to point or area data) would benefit from a combination of raster and vector mode algorithms. Line simplification and smoothing would take place in the vector mode, while rep-resenting the proximity of lines to neighbouring features and therefore pinning down any potential overlap would occur in the raster mode. Li and Openshaw (1992) have also proposed a line generalisation algorithm which, as the authors observed, worked better when using a hybrid raster–vector method; the vector approach maximised the smoothness of the lines, while the raster mode allowed the algorithm to operate more rapidly.

The different generalisation operators can be considered to be an attempt to move away from the subjectivity of manual generalisation, towards *objec-tive* or *numerical* generalisation. The aim of these operators is that individual preferences or prejudices should not influence the results. However, even when the same type of operator is used to generalise the data (e.g. line simplification), results can differ. This comes about because, for each type of operator, more than one algorithm may have been developed. Even if algo-rithms perform the same kind of task (e.g. all line simplification algorithms reduce the number of points used to represent a line), the effects that each of these algorithms can cause in the geometric properties of the generalised data can differ. McMaster (1987) shows how the geometric properties of the data, such as vector displacement, vary for different line simplification algorithms. The sequential application of the algorithms from feature to feature (for example, a simplification algorithm followed by a smoothing routine) may also produce different results, according to the order in which the features to be generalised are encountered and the sequence in which the algorithms are applied. These and other issues related to the development of generalisation algorithms are discussed in the next section.

1.3.1 The algorithmic approach

Generalisation algorithms constitute the building blocks of the automation process. Most raster-based generalisation operators use some type of moving window to filter or smooth regions of an image (Müller, 1991b). An early example of generalising within the raster format was presented by Monmonier (1983) on the generalisation of land-use maps, such as the application of geometric filters for bridging and deleting gaps in the data. More recently, Schylberg (1993) has developed a methodology for the generalisation of cartographic data in a raster environment. His prototype system could handle the generalisation operators of deletion, amalgamation and simplification.

In all of the algorithms developed for vector-mode generalisation, a strong emphasis has been placed on *line* generalisation. There are two main reasons for this. First, 80 per cent of the features found on a typical medium-scale topographic map consist of lines (Müller, 1991b). Second, automated generalisation of line features (together with feature selection) is a problem of lower complexity compared with those that involve other features such as areas (Weibel, 1986). Only points are easier to generalise than lines. According to Brassel and Weibel (1988), point features offer the greatest freedom with respect to their manipulation and therefore they are the map elements most easily handled in generalisation. Even so, sophisticated approaches can be necessary, such as that proposed by Mackaness and Fisher (1987) for the displacement of point symbols, constructed around a knowledge-based system.

Of the research on line generalisation, most work has concentrated on line *simplification*. Line simplification algorithms (also called data reduction) eliminate excess or redundant points on a digitised line. According to Robert McMaster, a method of distinguishing the different line simplification algorithms is based on the size of the search region used in point selection: 'While certain simplification routines deal with only the immediate "coordinate neighbours", others search extended sections of the line, or with a global approach, the entire line' (McMaster, 1987, p. 333).

The emphasis on the development of this kind of line simplification algorithm can be explained by the tendency of using a series of points to represent lines in vector-based GIS. Polylines (the technical term for using points to make up a line) are essentially a list of coordinate pairs and therefore the only method of simplifying the lines is by eliminating some of the points. Recently, some researchers have suggested alternative kinds of representation that would be better at modelling entire shapes, such as a meander in a river.

Fritsch and Lagrange (1995) and Plazanet *et al.* (1995) propose, instead of polylines, the use of representations developed from spectral tools such as Fourier series and wavelet decomposition. These representations are based on curvature and allow line details to be isolated in the frequency domain, and to be processed according to magnitude. The authors hoped that these new representations would remove the constraint imposed by the polyline representation and would allow new kinds of generalisation algorithms to be developed that would complement the classical algorithms.

One of the most important and widely used line simplification algorithms has been the Douglas–Peucker algorithm (Douglas and Peucker, 1973). This algorithm eliminates points in a line according to tolerance bands (i.e. points are eliminated if they lie within a certain given tolerance) using a global approach. Many GIS and mapping packages (ARC/INFO, LASER-SCAN, GIMMS and MAPICS to cite a few) have this algorithm as a standard line simplification routine, and it is often described in GIS-related textbooks and tutorials. Twenty years after it was first developed, researchers still propose ways of improving the Douglas–Peucker algorithm. Hershberger and Snoeyink (1992) have described methods of speeding it up, and Cromley (1992b) proposes some variations to the algorithm that make it computationally faster while still retaining fewer points. In addition, when new algorithms are developed their performance is often compared with the Douglas–Peucker algorithm. This is the case for the new algorithms proposed by Thapa (1988), Li and Openshaw (1992), and Visvalingam and Whyatt (1993). The Douglas–Peucker algorithm was chosen for the research covered in this book and was compared with the results from manual generalisation methods. An evaluation of the advantages and disadvantages of the Douglas–Peucker algorithm compared with other line simplification algorithms is described in section 4.6.

Further examples of the automation of line generalisation are line smoothing, line enhancement and line displacement algorithms. While simplification algorithms retain or eliminate points, smoothing algorithms only *displace* points in an attempt to give the line a smoother appearance through the reduction of the line sinuosity. In its simplest form this could be achieved by calculating a moving weighted average along a line. Simplification and smoothing algorithms often work in tandem. If a line simplification algorithm creates a line that is too spiky, then the line might need to be smoothed afterwards. The *enhancement* of lines entails the addition of detail to the simplified line. This can be done using fractal techniques, such as in the case of the algorithm proposed by Dutton (1981).

Despite its importance, *displacement* (of lines and areas) is one of the aspects of automated generalisation that has been least touched upon, due to its complexity. With displacement algorithms features can no longer be treated in isolation (as with the algorithms discussed previously), but proximity and adjacency relations also need to be evaluated. In other words, displacement algorithms are *context-dependent* generalisation operators. One of the few researchers working in this area has been Mark Monmonier, who has described it as 'the most enigmatic aspect of cartographic generalisation' (Monmonier, 1987, p. 25). Monmonier (1989) proposed a strategy for feature displacement which he called 'interpolated generalisation', whereby generalisation of a map – for example, at a scale of 1 : 250 000 – could be guided by two digital maps, one at a larger scale and another at a smaller scale (e.g. one at 1 : 100 000 and the other at 1 : 2 million). However, his strategy only dealt with features in pairs; and, often, in order to resolve a conflict between two or more features, it is necessary to take account of a larger number of features in a collective way. As features are displaced, not only do new conflicts arise, but the overall topology and relative position between features has to be maintained: the further away from the region of conflict, the less the displacement has to be in order to smooth the transition between displaced and not-displaced features. Mackaness (1994) calls this the 'displacement ripple effect'.

Mackaness (1994) proposes a new algorithm to take account of this 'ripple effect'. However, it treats lines as a sequence of points; in other words, the algorithm is essentially a point displacement algorithm. The algorithm is also restricted by the fact that it can only handle radial displacements and these frequently cause shape distortion. Moreover, displacement is not the only solution for a spatial conflict, and a comprehensive feature displacement approach will need to couple displacement algorithms with other algorithms. For example, the solution of a particular spatial conflict (such as the coalescence or overlapping of features) might require some objects to be eliminated and others aggregated before displacement is carried out.

The more sophisticated displacement system being developed by Anne Ruas does take into account the integration of displacement algorithms with other generalisation operators. Ruas (1995a) and Ruas and Plazanet (1997) describe the computations needed in order to determine which objects are overlapping or are too close to one another as a result of a previous transformation. In order to represent proximity between non-connected objects, the displacement system uses Delaunay triangulation to link an object with a set of neighbouring objects; that is, the triangulation creates a network of spatial relations between geographical features. The Delaunay triangulation is used not only to compute the 'proximity relations', but also to compute the necessary displacement vectors, and to keep a record of all the displacements carried out. The system can handle the displacement of different feature types such as roads and buildings for medium-scale transformations (between 1 : 15 000 and 1 : 50 000 when most displacement occurs) and is being implemented on the STRATÈGE platform (an object-oriented GIS devoted to contextual generalisation – see section 1.3.2). Although it cannot yet cover the entire range of possible feature conflicts, it is a step closer in developing a system that can deal with feature displacement in a comprehensive way.

Some of the generalisation algorithms applied to areas have been essentially line generalisation algorithms, the 'lines' being the outline of the areal features.

There are problems with this approach in that line algorithms are often inappropriate for dealing with areas, as they 'operate relatively locally and do not consider the area's overall shape' (Brassel and Weibel, 1988, p. 238). Müller and Wang (1992) approach all aspects of the generalisation of area features in the vector mode. The authors propose algorithms dedicated to areas that allow the elimination, selection, aggregation, displacement, smoothing, reduction, expansion and contraction of area patches. Their system checks for topological distortions, which may have been introduced due to the expansion and contraction of the areas.

As with areas, the generalisation of surfaces has sometimes been dealt with inappropriately, such as in the case of smoothing contours as if they were lines. Brassel and Weibel (1988) suggest that proper strategies should not be limited only to minor scale reduction, but should instead address individual landforms for elimination and simplification. Most work done on the generalisation of surfaces has been concerned with topographic surfaces. For example, Weibel (1992) proposes a strategy for terrain generalisation that is adaptive to different terrain types, scales and map purposes.

Each of the generalisation algorithms described above contributes to the automation of a *particular* aspect of the generalisation process. However, in order to attempt to automate the generalisation process as a *whole*, the many computer algorithms need to be linked together in a logical structure (McMaster, 1989). Also, the generalisation of the different features (sometimes stored in different layers in a GIS) need to be generalised in parallel, similar to the practice of a manual cartographer, rather than sequentially. Future progress in automated generalisation therefore relies on the process being 'viewed holistically and not as a series of isolated independent steps' (McMaster, 1989, p. 5). Most algorithms in use today generalise individual features independently of their geographical context (i.e. are 'context-independent'). The challenge lies in the development of algorithms that are guided by the geographical context, and in improving procedural knowledge of generalisation to know where, when and why certain generalisation operators and algorithms should be applied.

1.3.2 Knowledge-based systems

It is the complexity of the generalisation process as a whole that has motivated much of the research effort to use the concepts and techniques of a knowledge-based approach. A crucial test of any computer program for automated generalisation has been its ability to reach a sensible conclusion in situations in which a variety of alternatives exist. For a program to be able to make such choices (in cases in which all possible alternatives have not been explicitly defined in advance), there is a need for some form of reasoning mechanism to be included. It has been this requirement for a reasoning capability that has made investigation of knowledge-based techniques a potentially fruitful approach.

The use of such techniques does not mean that existing methods and algorithms are entirely rejected; rather, they provide a useful base upon which an improved system can be constructed. Most knowledge-based approaches to generalisation have used various algorithms to carry out the mechanics of the generalisation, with the expert knowledge being reserved for higher-level executive decisions – 'when' and 'how' to generalise. Important examples of such systems are: CHANGE, developed at the University of Hannover; STRATÈGE, developed at the Cogit Laboratory of the Institut Géographique National; and MAGE, developed at the University of Glamorgan.

With an earliest start date of 1978 and continuing until the present, the Institute of Cartography at the University of Hannover has the longest running research programme in computer-assisted generalisation of topographic information. This work has brought about the development of a complex system for generalisation called CHANGE, a stable prototype system that runs at several institutes and state mapping agencies in Germany and also abroad (Grünreich et al., 1992; Grünreich, 1993). This system has concentrated on the generalisation of the German basic scale maps for the scale transition from 1 : 1000 to 1 : 5000, and from 1 : 5000 to 1 : 25 000. Algorithms have been worked out using conventional programming methods with approximately a five per cent input of knowledge-based techniques (Herbert and João, 1991). Some components of the system have been designed for the generalisation of buildings, principally relying on the use of threshold parameters that set the minimum building sizes or distances for the operations of simplification and aggregation. Other modules have been designed for the generalisation of traffic routes and rivers. These algorithms make use of parameters to set minimum sizes or widths, and have also been used for emphasising features or for the simplification of irregularities.

Currently, the development work is focused on linking CHANGE to the GIS projects ATKIS (the topographic information system) and ALKIS (the digital cadastral map system) of the German surveying and mapping administration. The thrust of the development of CHANGE has been on trying to reduce the amount of user intervention. Manual revision of the final graphic output is still necessary for quality control and corrections such as displacement of conflicting objects in complicated surroundings. A test by the Cartographic Institute of Catalonia estimates that with CHANGE only 50 per cent of the work can be done automatically (Baella et al., 1994).

More recently, the development of automated generalisation systems has been aided by object-oriented data models, which are more suited to the task (see section 1.4). STRATÈGE (Ruas and Plazanet, 1997) and MAGE (Bundy et al., 1995) are new prototype systems that use a combination of algorithms and knowledge-based techniques, implemented using object-oriented technology. What these two research systems have in common is that they have been developed specifically to deal with more holistic or contextual aspects of generalisation; that is, to consider all of the features on the map simultaneously, and the interaction between them. This is more ambitious than previous attempts at automation, which dealt only with features in isolation. In order to achieve this, both systems need to determine proximity and adjacency relations, and detect spatial conflicts (such as overlapping features). Although distinct in their individual approaches, both systems tackle this by using Delaunay triangulation in order to connect the different features and therefore implicitly store object-topology. This data structure can then be used directly to perform complex context-dependent generalisation such as displacement and merging.

On the whole, all of the above systems have been very successful, but only within their restricted domains: they have generally concentrated on a particular scale transformation, or a specified data set, or a particular set of generalisation operations – or they still rely to a large extent on user intervention. For this reason, they are still limited in scope. All of the systems described above are also only prototypes and have mainly been developed for research purposes. In contrast to the performance of these systems, GIS presently available on the market offer fewer capabilities for database generalisation, and there is clearly an absence of a fully knowledge-based generalisation systems functioning within commercial GIS (for a description of the generalisation functions available in commercial GIS, see section 1.3.3).

The main reason for this absence has been because those who see knowledge-based systems as a potential solution come up against the difficulty of formulating rules for generalisation that are generally applicable (Buttenfield and McMaster, 1991). In theory, a knowledge-based approach requires that our knowledge of the generalisation process can be *formalised* into a chain of reasoning paths, each leading to a particular decision or procedure for generalisation to take place (Müller, 1991b). However, in practice, there has been little agreement as to how generalisation should be performed; that is, there is a lack of *procedural* knowledge.

In the cases in which generalisation laws have been developed, they place emphasis only on individual aspects of the generalisation process. This is the case of Töpfer and Pillewizer's (1966) 'Radical Law of Cartographic Generalisation', which calculates the *number* of objects that can be selected for depiction at a smaller scale (see section 2.2.1). However, this law only provides a quantitative measure of *how many* features: it neglects the crucial question of *which* features. In addition, the law does not address the occurrence of geometric alterations with scale change, such as in the case of the amalgamation of points to an area (McMaster, 1989). This lack of a unified theory for generalisation is identified by Fisher and Mackaness (1987) as a major obstacle to automating generalisation using knowledge-based techniques.

Table 1.2   Ordnance Survey guidelines for the depiction of water detail for the 1 : 10 000 scale map.

CANAL
5201. Canals are water detail and will be shown by double firm lines subject to minimum clearance between parallel lines (0.64 mm).

5202. Where the side of a building is coincident with the side of a canal, the side of the building will be shown as outline detail and that portion of the canal omitted as water detail.

5203. Disused canals which are dry and overgrown will be shown as outline detail.

RIVER, STREAM
5629. Rivers and streams will be shown as water detail. If the feature is less than minimum width it will be shown by a single line positioned centrally to the feature.

5630. Where a stream changes from a double to a single line it will be tapered. A stream which frequently changes from a double to a single line within a sheet will be generalised according to the major position but the situation on adjoining sheets will always be taken into account. Comparison across a minimum of 3 sheets (if available) will be made.

5631. In *basic* areas[a] rivers and streams will be shown as a double line unless their internal width is 0.64 mm or more at scale.

5632. Man-made sides of rivers or streams will be shown as outline detail.

5633. To avoid the double line representation being mistaken for two single streams, the application of flow arrows will normally clarify the situation, but where the feature widens to form the likeness of a pool that part of the stream will be shown as a pond.

INLAND WATER
General
5431. Inland water includes lakes, inland lochs and lagoons, natural reservoirs, ponds and any areas described 'Water' on the 1 : 1250 or 1 : 2500 maps.

5432. The natural sides of inland water features will be shown as water detail in accentuated gauge line (0.25 mm); man-made sides, e.g. dams, banks, weirs, will be shown as outline detail in normal gauge line (0.18 mm). Where a natural pond is coincident with a fence the water detail will be shown as complete and the coincident part of the fence omitted. However, where the fence represents the casing of a classified road it will be retained and the coincident water detail omitted.

Ponds
5434. Ponds smaller than 1 mm² will be omitted unless they are in an area where there is little other water detail, in which case they will be enlarged to the minimum size (i.e. 1 mm²).

5435. Ponds which are less than 1 mm wide but at least 2 mm long will be enlarged to 1 mm wide if space permits.

Table 1.2   (cont)

LEVELLING INFORMATION
Water level
5465. The winter level of lakes and inland lochs, and the top water level of natural
reservoirs, i.e. those terminated by dams, etc., will be shown provided that the area of
water is more than 0.25 km^2 in *derived* areas[a] or more than 0.50 km^2 in *basic* areas.
Where the dividing line between *derived* and *basic* runs through a lake, loch, reservoir
the *derived* area limitation (0.25 km^2) will be used.

Source: Ordnance Survey, 1989.
[a] *Derived* areas in a 1 : 10 000 scale map derive from the larger scales 1 : 1250 or
1 : 2500; while *basic* areas have been drawn directly at the scale 1 : 10 000.

Where explicit rules for generalisation have been created, they often relate
to the practices of a particular organisation, such as the rules for Canadian
national atlas production, as described by Richardson (1989), rather than universal rules for all countries and organisations. In addition, another problem
arises in that these specific rules are meant to be interpreted *by cartographers*
rather than by a computer. Some Ordnance Survey guidelines for the depiction of water features for the 1 : 10 000 scale map are set out in Table 1.2.
These are guidelines rather than strict rules, as they may be applied slightly
differently in different situations with the help of a little experience and 'cartographic appreciation' (Paul Chesterfield, OS, 1993, personal communication).
The guidelines shown in Table 1.2 do not necessarily relate to scale change
modification but, rather, give instructions for the depiction of features at the
scale of 1 : 10 000. For these guidelines to be transformed into generalisation
rules, they would need to be translated into particular generalisation operators, following a procedure similar to that described by McMaster (1991) for
the United States Defense Mapping Agency.

OSGEN, another classic knowledge-based generalisation system, illustrates
the difficulty of defining specific rules. The OSGEN system, developed exclusively as an advisory system, was intended to test the rules used by the Ordnance Survey (OS) cartographers and surveyors. The system, developed
between 1988 and 1989, coded the rules (about 300 of them) using an expert
system shell, and then assessed them for internal consistency and completeness
(Robinson and Zaltash, 1989). Because OSGEN was meant to be used entirely
in an advisory role, it had no algorithms to carry out the generalisation. The
generalisation would still be done manually by the cartographers, who would
ask for the advice of the expert system in difficult situations. OSGEN advised
on symbolisation, aggregation, displacement, simplification, enhancement and
elimination in the production of maps at a scale of 1 : 10 000 derived from
scales of 1 : 1250 and 1 : 2500. The system used a series of menus and prompts

to ask relevant questions about selected features, and, on the basis of the answers to these questions, a recommendation for action was made. Importantly, the user was able to ask the system why a particular question was being posed or why a particular conclusion had been reached.

Although only intended as an advisory system, OSGEN still ran into difficulty in the development of the project's rule base, due to the nature of the rules that could be formulated. Because of the multiplicity of possible arrangements of data, many of the rules were found to be 'fuzzy' in nature and would need interpretation by the user of the system. As a result of this research project, it transpired that there were also a significant number of unwritten rules in use, depending on the particular context. Also, some rules that appeared to be 'hard' turned out to be more flexible when subjective factors were considered. The OSGEN project showed the extent to which generalisation, even by a national mapping agency such as the Ordnance Survey, with its formal set of guidelines, still relies on judgements being made by those supervising the generalisation process.

The main problems in automating generalisation have been the formalisation of the rules (see Buttenfield and McMaster, 1991) and the formalisation of a process that has remained essentially highly intuitive and subjective. Cartographic knowledge is difficult to pin down. However, even if this challenge was surmounted, an automated generalisation system would still need to be capable of dealing with a diversity of generalisation methods, the many different map purposes and map scales, and, above all, the huge variation in the characteristics and spatial distribution of the data that can be involved.

Successful generalisation has to be a combination of both local and global factors, independently of whether it is carried out manually (by a person) or automatically (by algorithms and/or knowledge-based techniques). According to Mackaness and Scott (1988), a cartographer performing generalisation needs to make simultaneous decisions about groups of features (local level) and about the layout of the map as a whole (global level). Local relationships can affect the application of global generalisation rules. For example, the generalisation of population centres does not depend solely on total population, since local density can influence whether or not small towns are retained. And roads, railways and houses that have been generalised should preserve the correct relationship between the three entity types.

To overcome these rather daunting problems, Müller and Wang (1992) suggest that there are three main strategies for developing a truly flexible automated generalisation system. The most ambitious strategy is a 'top-down approach'. All cartographic objects are generalised simultaneously and, most importantly, the attributes of the objects and relationships between them are retained across the scales. Less comprehensive is a 'bottom-up approach' in which each particular class of features is dealt with one at a time. Classes may be based on geometry or semantics. This would be especially appropriate for special-purpose maps (e.g. hydrographic). Only later would a more holistic approach be tackled. Finally, a simpler but more pragmatic proposal is to provide a 'toolkit' of generalisation functions (e.g. selection/elimination, simplification, etc.) which are in an algorithmic form and could be called up where necessary. This sort of approach would need to rely on an extensive amount of input from the user, but over time more and more of the system could be automated.

1.3.3 The return to human control

The extent to which generalisation could be carried out automatically and the necessary involvement of skilled manual intervention is (and has long been – see Rhind, 1973) a matter of debate. While researchers in the past have struggled with the *full* automation of the generalisation process ('batch generalisation'), the way forward now seems to be to rely on some form of human control ('interactive generalisation'), even if only as a transitional approach to a comprehensive automated system. The suggestion that the role of the user will always continue to be crucial in some form or another has been prompted by the fact that existing knowledge-based systems have not succeeded in automating the generalisation process, because of its complexity. The main challenge now is to determine where the boundary between manual intervention and automated techniques lies.

Weibel and Buttenfield (1988) were the first to suggest that the automation of generalisation should be subdivided between the user and the computer according to problem-solving potential. The computer would implement some tasks that it was good at solving, but would rely on the user for control and knowledge. The authors expected the user's knowledge to be gradually extracted and formalised, leading over time to the development of a full-scale expert system approach. The authors labelled this approach *'amplified intelligence'*, as opposed to *artificial intelligence.* The authors considered that the strategy of 'amplified intelligence' could solve the problem of automating generalisation by using existing cartographic software as a 'workbench' at which knowledge could be increasingly acquired.

Weibel (1991) and Beard (1991b) have further developed the argument for an 'amplified intelligence' approach. The objectives of this alternative approach for the automation of generalisation are 'first to overcome the weakness of algorithmic approaches by incorporation of knowledge into generalisation systems, and second to eliminate the deficiencies of knowledge engineering strategies by providing a structured approach to knowledge acquisition' (Weibel, 1991, p. 177). This approach therefore lies between the algorithmic and the knowledge-based approaches described previously. The inclusion of human intelligence in the automated map generalisation process is suggested in these cases where the automation of generalisation *fails*, and the

user intervention is needed to help the system to perform the complex generalisation procedure. Weibel (1991) argued that this type of strategy was a transitional one and would only last until the system had learnt and incorporated
all the human expertise.

In addition to human intervention, Weibel *et al.* (1995) have suggested that
the use of computational intelligence techniques, such as genetic algorithms
and artificial neural networks, can also help to overcome the deficiencies of
traditional knowledge-based techniques such as expert systems. The reason for
this is that, unlike expert systems, computational intelligence techniques do
not need a full set of generalisation rules, as they can 'learn' from sets of
positive or negative examples (i.e. implicitly represented knowledge). There is
an obvious advantage in a system that can remember solutions to different
types of problem. The advantage is even greater if it can abstract from these
solutions potential strategies in order to deal with similar, but not identical,
problems in the future, or if it can extract new rules that are implicit in a
solution. This sort of capability reduces the amount of initial development of a
program that would otherwise be needed. The range of possibilities that exist
is too large for a system to be constructed in advance with specific rules to
deal with every possible situation. Nor is this necessarily desirable, as it may
result in systems containing many rules that are rarely, if ever, used.

Computational intelligence techniques are, however, computationally intensive and require the existence of clear criteria as to what a good generalised
map is: 'Probably the most important research problem, and thus the key to
further progress in the use of computational intelligence techniques in generalisation is, however, the development of effective measures and methods for
evaluating the quality of generalisation results' (Weibel *et al.*, 1995, p. 154). A
final end state also needs some consideration, as João *et al.* (1993) point out, as
there may well be several different though equally good solutions to the user's
preferences. It is also necessary to determine how best to handle 'generalisation failure' in cases in which the system cannot meet the specified generalisation objectives.

Irrespective of the success in developing fully automated generalisation
systems, it can be argued that user intervention will always be needed (João,
1991). There are different types of user, some with more experience than
others, and some doing very different work from others. This serves to emphasise the need for flexibility in accommodating different user needs, and to allow
users interactive control in specifying factors such as map purpose or type, and
scale. Beard (1991a), for example, argues that a flexible generalisation system
must permit users to create a wide variety of different views of an original data
set, and not be restricted to the production of traditional general-purpose
maps or a few pre-defined map products. So, even if in some cases fully automatic production was possible in a pre-defined map production environment
(as in the case of national atlas mapping), user interaction would always be
required in the cases in which users are likely to choose from amongst many

different scales and different final results. Much of the subjectivity involved in designing or generalising maps can come from differing user preferences, so emphasis on flexibility to accommodate these, and on allowing for user intervention in appropriate circumstances, will be necessary to maintain the diverse character of the processes to be modelled.

Users have a key role in deciding the purpose and the extent of generalisation, and in evaluating the final result. An automated system should be able to support their preferences, and should ideally be adaptable to the methods of particular organisations. According to this argument, an effective automated generalisation system would require a flexible and friendly enough user interface to deal with a wide range of users and queries. While this is crucial conceptually, it continues to prove difficult to implement. Petchenik (1983), for instance, has pointed out how complex the hypothetical user can be, because even if the responses to certain directed tasks are known, little might be known about the questions the user might ask spontaneously.

The difficulty in implementing such a concept possibly justifies the fact that only recently has a commercial system been developed based on these principles. Introduced in the spring of 1993, Intergraph's MGE Map Generalizer (MGMG) was the first commercially available system for *interactively* controlled cartographic generalisation. This system relies on a toolkit of generalisation functions which are programmed in algorithmic form and can be individually called and executed by the user. The system applies in practice the principle of 'amplified intelligence': 'The interactive approach gives the user the most flexibility and control on what to generalise, how to generalise, and how much to generalise' (Lee, 1995, p. 219).

The available generalisation tools of the MGE Map Generalizer, designed for vector-based processes, were until recently the most extensive of any commercial GIS. The set of generalisation operators includes the elimination of features (based on area or length, for example), smoothing and simplification of lines, collapse of features (e.g. of areas into points or lines, and of a line into a point), aggregation (both of point and areal features), classification, simplification of point clusters (typification), and some aesthetic refinements such as symbol rotation. Although the system contains an operator that identifies potential displacements needed, it does not actually displace features (Weibel and Ehrliholzer, 1995). This, together with the lack of tools for exaggeration, can be considered as limitations of the early versions. Displacement and exaggeration were probably excluded initially because of the complexity of these generalisation operations.

To date, the only other fully operational commercial system for automated generalisation is the object-oriented mapping system LAMPS2, launched by Laser-Scan in the summer of 1995 (Woodsford, 1995). This system has the advantage of being object-oriented and possessing good topology management. Data models for generalisation are really only effective if they can record spatial relationships between features, and for this object-oriented data models are better than layer-based models. The generalisation functions available in LAMPS2 are extensive and include simplification, typification, aggregation, classification, collapsing, exaggeration, refinement and, most importantly, displacement. LAMPS2 can resolve conflicts between closely neighbouring objects. The distance and direction of displacement are determined by priority, width, displacement factor and a line sensitisation factor, defined class by class (Laser-Scan, 1997).

Both the Map Generalizer and LAMPS2 are interactive in that they depend on the user to select the geographical features to be generalised, to select the generalisation operators, to choose the particular algorithms, to input the generalisation parameters, and finally to accept or reject the result on the basis of visual inspection. The decision-making process regarding when, where and how much should be generalised is therefore left to the user. As a consequence of this, the developers of the Map Generalizer assume that the users are trained in cartography and have a thorough knowledge of manual generalisation (Lee, 1992): 'This approach is in fact an electronic analogue to manual generalisation with a high demand placed on user–machine interaction' (Müller and Wang, 1992, p. 138).

It seems likely that, if the process is to have satisfactory results, the GIS user will always need to play an important role in GIS generalisation. From the specification of user's needs to the solving of complex data conflicts and the evaluation of the final result, the user should be given room to increase or control the system's performance. It is vital therefore that the interaction of the user with the GIS be as efficient and simple as possible.

At present, commercial generalisation systems still lack any help to the user in terms of what generalisation algorithms to use, which generalisation parameters to select and in which order to apply the algorithms. The 'undo' command of the Map Generalizer, which returns the display to the previous step, does make it easier for the user to experiment with generalisation parameters and commands, but the procedure still requires considerable trial and error. Users of existing GIS generalisation systems are supposed to adjust the different generalisation parameters until the end result *looks good*. This, of course, assumes that all users know what they want and what the end result should look like.

An exclusive reliance on the visual inspection of results to determine the quality of the generalisation process is a severe limitation. While existing GIS generalisation systems can vary in their generalisation capabilities, they all have very poor facilities to analyse and control the changes made to the data due to the generalisation. According to Lee (1992), future releases of the Map Generalizer will provide some capabilities for statistical measurements to analyse the original data and evaluate the generalised map. However, these measurements will be intended mainly to derive generalisation parameters and to load parameter files, rather than to analyse the impact of generalisation on data quality. Independently of what the ultimate user interface would look like, the interface should make clear to the user the cause and effect between method used and result obtained. The identification of the relationship between generalisation algorithms and generalisation effects for different classes of features and for different scale transitions is the key element within the system proposed in this book, in section 7.3.

1.4 From multiple representations to scale-independent GIS

As long as the problem of automating generalisation remains unresolved, users of current GIS are likely to continue to store maps of different scales independently. In other words, *multiple representations* of the same reality will coexist in the same database. This sort of scale-dependent database is common in most mapping agencies at present. However, eventually it is hoped that systems of multiple representations will be replaced by a more advanced scheme. The ultimate goal of the GIS community and of mapping agencies would be to use a knowledge-based system and generalisation algorithms to produce a *scale-independent GIS* (also sometimes called scaleless or scale-free database). Instead of having to store multiple representations of different-scale maps (produced either from manual generalisation or automated generalisation methods), only one single large-scale database would be needed (with data collection and revision set at the level of precision for which the data were captured; Müller, 1991b): 'Data collected from ground surveys may be called scale independent, since the notion of scale only appears when those data are being transcribed, for analytical or representation purposes, onto a space which is smaller (or larger) than the original surveyed space' (Müller, 1991b, p. 471).

From this large-scale database, the different-scale maps could be produced (i.e. generalised) in real time as and when they were required. In a scale-independent GIS, if a feature such as Heathrow airport was selected, for example, at the large display scale detail such as airport runways, buildings, fuel tanks and parking lots would all be present, but as the user zoomed out only the generalised representation of the airport runways might appear at any display scale chosen by the user; say, 1 : 45 887. Zooming out still further (to a display scale of 1 : 667 650), the airport location would be displayed as an aeroplane symbol.

Müller (1991b) considered that there are three main arguments for the development of a scale-independent GIS: first, to avoid duplication in storage; second, to allow the production of flexible scale-dependent outputs (as any scale could theoretically be derived from source data, and not just the conventional steps of 1 : 10 000, 1 : 50 000, etc.); and, third, to ensure consistency and integrity between the various scale outputs, and to facilitate updating procedures. Beard (1987, p. 211) also pointed out that to create, store and maintain a single large-scale database and to derive all smaller-scale data sets from it 'should be less time consuming, less costly and potentially less error prone'. The main problems of the creation of a single large-scale database (assuming that the problems of automatic generalisation have been resolved) might be the unwieldy size of the files and the excessive processing time (which would delay real-time applications), and the inadequate treatment of scale-related variability within a file (Beard, 1987). The first disadvantage might not be a major problem in the future, if the current growth in computer processing power continues. The second problem would be helped by using feature-based or object-oriented data models.

According to Guptill (1989), of the United States Geological Survey, future database designs need to be produced that do not preclude users from creating seamless and scaleless versions. He considered that adopting a feature-based, object-oriented data model (e.g. in mapping agencies) was a positive step towards this goal. This was also the opinion of Müller (1991b), who argued that, in order to support multiple representations from one single source, a feature-based data structure, combined with generalisation operators and decision rules, would be needed. If a database for generalisation is feature-based, then an object-oriented programming approach to generalisation would be the ideal solution. In an object-oriented environment, procedures are linked to the object itself (e.g. a coastline) and objects execute their methods (i.e. generalisation) directly in response to received messages (Egenhofer and Frank, 1989). Using such an approach the various representations, at different scales, of a single feature may be linked so that they 'inherit' common characteristics. Any update of one of the representations could then appear concurrently across all map-scale layers in the database (Müller, 1991b). However, in order for this to work effectively, it has to be assumed that a way can be found to match different representations of the same real-world objects and, as Devogele *et al.* (1997) have shown, this is not always straightforward.

The potential of a feature-based approach to show different views of a single database has been demonstrated by the 'adaptive scaling' display within the Smallworld GIS. This has a few of the characteristics of a scale-independent database. The two main components of this adaptive scaling are the facility to change the visibility of the different features (i.e. features can be set 'on' or 'off' for ranges of display scales) and the facility to change the style of the representation of the features (e.g. a city could be displayed as a point at the scale 1 : 1 million, but with its outline at the scale 1 : 50 000). In this GIS, if the scale-dependent option is chosen, then every time the scale is changed (either explicitly by requesting a particular scale, or by zooming in or out), the features on the screen are drawn using the style defined for the display scale in the current view scale range (Smallworld Systems, 1994).

The advantage of using object-oriented data models is reflected in the success of recent generalisation systems with object-oriented data representation, such as STRATÈGE and MAGE (see section 1.3.2), and how they achieve complex context-dependent generalisation functions such as displacement. Schlegel and Weibel (1995) proved how an unsuitable data model could be an impediment in the development of generalisation tools. The authors developed a prototype system, called GenTools, which extended the basic functionality of ARC/INFO for generalisation. In trying to develop new generalisation tools for this general-purpose GIS, the authors found that one of the major problems was the restrictions imposed by the layer-based data model of ARC/INFO. This data model did not ensure a good interplay of the different feature classes, was not rich enough to support complex operators such as displacement and amalgamation, and topological inconsistencies could not be easily checked.

As a first step towards a genuine scale-independent database, Jones and Abraham (1986) proposed a sophisticated version of a *pseudo*-scale-independent database. The authors proposed a hierarchically structured database in which different scale-dependent layers of representation could be accessed without duplication of data. Assuming that the points selected for small-scale representations were always a subset of those used in larger-scale representations, then the technique could reduce the amount of multiple line storage, while avoiding the excess requirements of single, large-scale storage (Müller, 1991b). The organisation of the data into an hierarchical structure allowed for the selection and retrieval of information considered relevant at a particular level of resolution. However, this type of Binary Line Generalisation tree (BLG-tree), described in detail by van Oosterom (1993), is dependent on the algorithms used to allocate the data for the different levels. Jones and Abraham (1986), for example, used the Douglas–Peucker algorithm, and therefore the quality of the resulting 'scale-independent' database was related to the quality of the algorithm's performance in rank ordering the points. Another criticism was given by van Oosterom and Schenkelaars (1995), who pointed out that these structures are used for the generalisation techniques of simplification and selection only, and that other types of generalisation transformations – such as combination, symbolisation, exaggeration and displacement – could not be performed within the framework.

A single-scale database approach requires considerations far beyond the development of automated generalisation algorithms. Coupled with the notion of the development of scale-independent databases is the notion of *seamless* databases. A seamless database would entail an ability to query, display, retrieve or otherwise move across the contents of a large spatial database without any limitations imposed by the spatial extent of the data (Guptill, 1989). For example, if desired, the River Thames could be structured as a complete entity, rather than as portions of it from separate map sheets. This would not necessarily be straightforward. Smith (1989) pointed out that even a single 1 : 25 000 urban sheet covers the area of 800 1 : 1250 plans and has 1540 internal edges. Each of these would need to be matched and checked. Therefore, assiduous edge-matching across map sheets would be necessary in order to create a scale-independent and seamless database. However, this would only be necessary once, and from then on only updates would be required. An early example of a seamless database is the case of the Superplan services from the Ordnance Survey, which permit users to obtain a product for non-standard areas (see Sowton, 1991). In other words, users can create a map centred at a point of their choice without being restricted by conventional map sheet edges.

There is no doubt that the development of a scale-independent GIS would represent a quantum leap beyond the multiple-representations database, but it will be a difficult challenge: 'In the long term, the concept of a scale-free database is a Holy Grail to which we must aspire but reaching it is probably nirvana' (Rhind, 1990, p. 91). Considerable generalisation research is envisaged in order to overcome this challenge: 'Generalisation functionality is still severely lacking from today's GIS and digital cartographic systems. This lack of adequate generalisation functions may be regarded as one of the most serious impediments to a flexible and meaningful use of spatial data in GIS' (Weibel, 1995a, p. 260).

Despite the vast amount of research being carried out on developing automated generalisation, there is one particular topic that is rarely touched upon. While all researchers agree that generalisation influences spatial data, few have investigated the *effects of generalisation* on data sets collected at different scales. In the next chapter, I review the work done in the past to investigate the magnitude of generalisation effects on different types of geographical data. In most work, generalisation effects have only been observed *qualitatively* and there have been very few studies that actually evaluate the effects of generalisation *quantitatively*.

CHAPTER TWO

The consequences of generalisation: a review

Generalisation is important to GIS, not only because of the drive towards automating the process and the development of scale-independent databases, but also because generalisation can seriously affect data within a GIS, even more so than within paper maps. Despite this, there are almost no *quantitative* estimates of the magnitude of generalisation effects, and most published literature describes the consequences of generalisation only in qualitative terms. This book aims to provide such quantitative data.

More generally, the increasing use of GIS has brought profound worries about sources of error in maps, including those caused by generalisation effects. This is because 'GIS has the potential to dramatically increase both the magnitude and importance of errors in spatial databases' (Openshaw, 1989, p. 263). The magnitude of these errors surpasses those introduced by traditional cartographic manipulations of paper maps: 'The problems result from the power of GIS to routinely perform operations on cartographic data which traditionally would not have been done, or else performed only under special circumstances, because of the problems of scale, complexity, and feature generalisation that might be involved' (Openshaw, 1989, p. 263).

Unaware users can still be fooled by the *precision* of GIS manipulations and might wrongly assume that this is synonymous with *accuracy*. In addition, data contained in a GIS do not have a specific scale (as the user can easily change the scale by 'zooming in and out') and as a consequence the user can ignore the fact that, despite appearances to the contrary, all spatial data have a specific resolution associated with their original collection. As discussed in chapter 1, data are produced with variable levels of accuracy and resolution and, therefore, data are only appropriate to be used within a certain range of scales (Aronoff, 1989): 'If the data were obtained by digitising, their resolution is that of the digitised map. If they were obtained from imagery, their resolution is that of the pixel size of the imagery' (Goodchild, 1991, p. 114).

This difficulty with digital data – that it suggests higher accuracy than it should – is less of a problem with analogue maps. Openshaw (1989) explained that this was because in manual cartography many of the problems are 'visible', and skilled operators make the necessary adjustments and form professional judgements about data quality. But with a GIS, the operators are not necessarily so aware of all the limitations of the data and the problems become less visible. In GIS, 'the apparent ease by which data from different scales and qualities of map document, with different levels of innate accuracy, can be mixed, integrated, and manipulated totally disguises the likely reality of the situation' (Openshaw, 1989, p. 263). This can easily lead to the creation of spatial databases with unknown accuracy.

The problem of error is compounded with digital maps because, unlike their paper counterparts, their inaccuracy might not be explicitly stated. The manual cartographer goes to 'considerable pains not to let the map give an impression of an accuracy greater than the source material warrants' (Robinson et al., 1984, p. 134). Because a well-designed and well-produced map can have an authoritative appearance of truth and exactness, cartographers must make sure that they convey to the map reader a clear indication of the quality of the data employed in the map. In order to transmit the unreliability of their map, cartographers use symbols and the map legend. In geological maps, for example, it is common practice to show precise boundaries by continuous lines, but to change to a dashed symbol where the boundary is less well defined (Fisher, 1989). When appropriate, the cartographer can also include a statement in the legend concerning the accuracy of any item. As in the case of analogue maps, any digital database creator should also include such information and should indicate its quality to avoid misuse. Unfortunately, this is rarely the case and in extreme situations even basic information regarding the source scale might be missing!

This is exacerbated by the fact that few users have the luxury of creating all their own digital data. Typically, data will be imported from a variety of sources, the number depending both on the data collection procedures employed in the country and the primary tasks for which they were intended. In such circumstances where data have been supplied externally, prior generalisation effects are embedded in the data. To quantify the effects of potential errors, it becomes essential to have access to information about the sources and character of the data, and the procedures to which they have been subjected (Chrisman, 1987).

GIS-based analyses of map data can also generate uncertainty that is uniquely due to the manipulations themselves. A notorious generalisation-related artefact, which results from handling generalised data in a GIS, is the problem of *sliver polygons*. The generation of these polygons is 'the most

common form of error in overlaid maps' (Chrisman, 1989, p. 526). Sliver polygons result from the overlaying of different digital maps (e.g. with different
amounts of generalisation) of the same geographical area. Differences in the
detail of the geographical features cause small unintended polygons to be
created. For example, in the case of an overlay of soil and land-use data,
boundaries that should be the same (e.g. caused by a river) might not match
up, producing tiny spurious areas. Despite the fact that they are small and that
they occupy a small proportion of total map area, large numbers of sliver
polygons 'create serious problems for topological data structures and should
be suppressed' (Goodchild, 1980b, p. 202). The creation of sliver polygons is
another manifestation of the generalisation problem within GIS, and exem-
plifies the difficulty of discerning which relationships are real and which are
artefacts of the data creation or transformation processes.

Other sources of potential uncertainty in cartographic data could only be
easily dealt with once GIS was introduced. Such is the case of the Modifiable
Areal Unit Problem (Openshaw, 1989). One of the implications of this aggregation problem, which is discussed in more detail in section 2.2, relates to the
fact that the statistics resulting from different agglomerations of a set of areal
units may vary considerably. According to Openshaw (1989), this problem
became of much greater operational importance, and stopped being an academic curiosity only when GIS freed the user from having to use a limited
number of fixed, usually 'official', sets of areas for geographical analysis.

From the above, it can be concluded that the uncertainty associated with
digital data can be considerably larger than the uncertainty inherent within
the traditional map products from which they are frequently derived. In addition, GIS map manipulations increase the magnitude and the importance of
errors in spatial databases, and they can also generate artefacts (such as sliver
polygons) which are caused by these manipulations. On top of all this, users of
digital geographical databases may still make unreasonable or inappropriate
assumptions about the accuracy of digital data.

While it is widely recognised that generalisation introduces error, the question arises as to whether this error is large enough to be important. However,
it is difficult to say beforehand exactly which generalisation effects are important and which ones are not, as the importance of these errors depends to a
large extent on the application of the data (Openshaw, 1989). Also, even if the
error due to generalisation is inferior to other sources of error (e.g. surveying),
it will still be important if, when combined with the other errors, the total
error becomes superior to a maximum admissible value. The generalisation
error may be 'the final straw' and the data may be discarded because of the
overall inaccuracy. Also, the types of error that contribute least to the total
error may be the simplest to reduce or correct (for example, reducing generalisation effects might be cheaper than surveying an area again). Therefore, if
the reduction of any particular error source reduces the total error to below a
certain threshold of acceptability, then the control of that particular error is

very much a valid exercise, irrespective of its magnitude.

This chapter reviews what has been covered in the literature in acknow-ledging, evaluating and measuring the effects of generalisation. As much as possible, generalisation effects are classified in relation to different gener-alisation methods and in relation to different feature types. However, gener-alisation effects are not the only source of map error. There are other types of error that are introduced when producing a map, ranging from surveying errors to paper deformation. This chapter starts by analysing all the known sources of map error and comparing these with the errors introduced by gen-eralisation.

2.1   Error associated with producing a map

During the map-making process there are a number of opportunities for the introduction of errors. The consequence of these errors will be, for example, that the measured distance between a house and a road intersection rep-resented on a map will differ from the true distance between these points. In practice, however, 'it is impossible to determine the true length of a line or the true area of a parcel because every measurement which is made, irrespective of the method used and the care taken, is encumbered with errors' (Maling, 1989, p. 84). The measured length on the map can therefore be only an approx-imation to its true value, and the less the errors in the map and in the mea-surement process are, the closer that measured length will be to that of the true value.

Positional accuracy determines how closely the position of discrete objects shown on a map or in a spatial database agree with their position on the ground. There are two main components of positional accuracy (Smith and Rhind, 1993):

- Absolute accuracy – how accurately the position of a feature is recorded relative to a reference framework such as the British National Grid.

- Relative accuracy – how accurately the features are represented in relation to each other. This relates to the topological consistency and the ability to measure accurately inter-point distances in the data or on the map.

This division is particularly valid in the case of error due to generalisation, in which the cartographer will frequently improve the relative accuracy of the map at the expense of the absolute accuracy. This approach is also useful in explaining the impact of systematic versus random errors. Assuming that gross errors have been eliminated, then the total positional error of a map or data-base can be considered as the sum of both systematic and random errors introduced during each stage of the production of the data (Bolstad *et al.*, 1990). Random errors will affect both the absolute and relative positional accuracy, while systematic errors will only affect the absolute accuracy. Sys-tematic errors introduce bias in the observations and therefore must be detected and corrected. However, even after all of the systematic errors have been removed, some variations in observations will still remain; these are the random errors, which are usually small in magnitude (Thapa and Bossler, 1992).

The accuracy of spatial databases goes beyond the issue of positional accuracy covered in this section, since it is also concerned with lineage, logical consistency, attribute accuracy, completeness and temporal accuracy (see NCDCDS, 1988). In addition to these errors, there are others that can affect the measurement process itself; in other words, errors that are considered likely to influence the accuracy of the measurement but are independent of the accuracy of the map. These are related, for example, to the way in which GIS carry out measurements on the data. These measurement errors are not investigated in this book because all of the measurements were made using the same GIS functions, and so, irrespective of the size of the error, the measurements remained comparable.

For this study, errors that can affect the positional accuracy of a digital map are classified into three main types: non-generalisation error associated with the production of the source map (from the initial survey, through compilation, to the map reproduction); error resulting from the digitising of the analogue map; and error due to generalisation. The positional accuracy of a digital map is first related to the error associated with the production of the original analogue map. Within this, it is relevant to distinguish between the production of basic scales and the production of derived maps. The successive stages in the production and revision of basic and derived maps, both in the case of traditional and modern map-making procedures, are illustrated in Figure 2.1. In some cases the procedures might be more complex or varied than those shown in Figure 2.1. The figure is therefore illustrative rather than a universal and definitive model. The traditional and recent map-making procedures are inter-linked because many maps held digitally were originally created by a traditional procedure (Fisher, 1991).

Different countries have different approaches to the production of map data. For example, the Portuguese national mapping agency still relies heavily on the traditional methods. In contrast, the Ordnance Survey has modernised to a much greater extent. This picture is complicated by the fact that many maps created traditionally are now updated using modern procedures. This updating process could cover only specific sections of the map, as some features change rarely (e.g. contours) while others (e.g. roads) can change every few years. This sometimes results in a particular map sheet being a 'patchwork quilt' of different map-making procedures. However, certain individual maps have recently been created from scratch using entirely modern methods.

Figure 2.1 is a simplification of the current situation, because it mainly reflects the old lithographic work and the old automated cartography (where the aim was to replicate manual processes by automated equivalents). Nowadays, some mapping agencies are progressing towards a more sophisticated

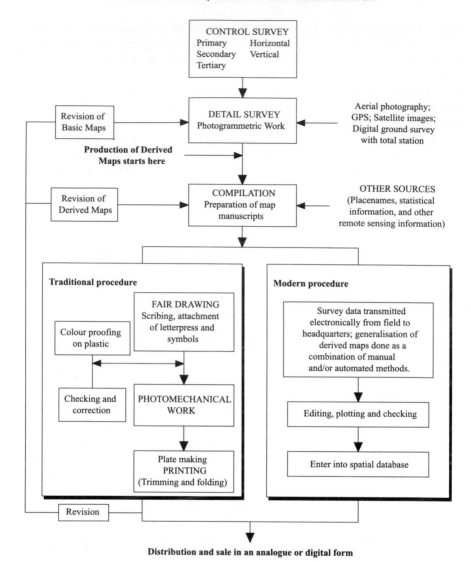

Figure 2.1 The main stages in the production and revision of basic and derived maps, both in the case of traditional and modern map-making procedures (based on Maling (1989) and on information from the Ordnance Survey).

database approach to spatial information. In this case, rather than individual digital map sheets, the whole surveyed area would be covered by a seamless database. As described in section 1.4, this would entail the very large task of edge-matching thousands of individual map sheets. The Superplan services of the Ordnance Survey are an example of a first step towards this database approach. Despite these advances, most maps extant (including digital maps and the maps used in this study) have been produced by conventional methods.

The survey of a country is normally created as a series of maps at the largest scale, known as *basic maps*, which are convenient for the majority of administrative, economic or defence requirements (Maling, 1989). The scale of the basic maps can vary for different countries, or even for different areas within the same country. For Great Britain, the scales 1 : 1250 and 1 : 2500 cover urban and lowland areas, while the scale 1 : 10000 covers all other areas such as mountains. In contrast, the basic scale map produced by the Portuguese national mapping agency is at the scale of 1 : 50000. Smaller-scale maps of the same country, which use the basic scale mapping as their source, are termed *derived maps*. Ground features such as rivers and roads are usually transferred to the maps by being plotted from aerial photography or by being derived from existing maps: 'Nowadays, it is exceptional for a new map to be plotted entirely from ground surveys' (Maling, 1989, p. 14).

The positional accuracy of any point on a map depends partly upon the accuracy of the survey and partly upon the accuracy with which points have been located by the plotting, the drawing and the printing processes (Maling, 1989). According to Maling (1989), the combined errors of the different stages of traditional map production shown in Figure 2.1 for any particular feature (plotting the control and the detail, compilation, fair drawing, reproduction and colour registration) can range from ± 0.42 mm to ± 0.73 mm.

The generalisation error is, however, 'very difficult to quantify because the amount of error introduced depends on the type of feature and also on the character (or complexity) of the features' (Thapa and Bossler, 1992, p. 838). As the basic maps get transformed via cartographic generalisation into derived maps, generalisation has a larger influence on the accuracy of the derived maps than for basic maps. However, even very large basic scale maps are generalised and are unable to show a complete record of the landscape, or to show all features true to scale. According to Harley (1975, p. 33) even a map from the Ordnance Survey 1 : 1250 series will, 'like any map, omit some details, generalise others, enlarge other features to a minimum size, and use conventional signs to codify much of the surface detail'.

The scale of the maps not only influences the degree of generalisation but also determines the relative importance of the different sources of error: 'The smaller the scale of the maps, the less is the importance to be attached to the errors in surveying and the greater is that attached to the errors of the graphical and reproduction stages. Thus most maps of scale 1 : 25000 or smaller which have been derived from modern control survey and photogrammetric plots have errors of drawing and reproduction which are much larger than those derived from other sources' (Maling, 1989, p. 157).

Independently of the error introduced by the map-making procedures, there are two main extraneous sources of error. One results from neglecting to take into account the deformations that are inherent in the representation of the curved surface of the Earth as a plane map. Depending on the map's projection, spheroid and datum, maps will suffer local scale changes. If local scale factors are ignored, then the resulting measurements – in other words, the conversion of map units (in millimetres) to ground units (in metres) – will be less accurate than they could be. The size of this error usually increases as the scale of the map decreases. The importance of this source of error depends on the area of the Earth's surface being mapped and on the projection or transformation used. Some projections introduce more distortion than others; for example, the Ordnance Survey National Grid was tailored to minimise such distortion for Great Britain.

The other source of error that can affect measurements made on a map is due to fluctuations in the size and shape of the map itself; that is, paper deformation. The regular or irregular alterations in the dimensions of the printed maps arise from variations in temperature and humidity during manufacture, use or storage of the maps (Maling, 1989). Paper is especially sensitive to changes in humidity and to a lesser extent temperature; but the dimensions of other materials used during traditional map-making processes, such as metal plates or plastic sheets, can also be affected by these factors. Most mapping agencies digitise from plastic originals because plastic materials are less susceptible to distortion than paper maps. Nevertheless, if maps were in a paper form before they were digitised, and no correction was applied, the paper deformation will be 'locked' into the digitised data. In this study, distortion due to paper deformation was minimised by transforming the maps accordingly (see section 3.3).

The deformation of the material affects the results of measurements as the map projection does: it changes the scale of the map locally. The distortion of maps due to paper deformation and map projection, when unavoidable, require that some form of correction is applied when the measurements are reduced to the ground dimensions (if they prove to be too large to be ignored). This deformation of the mapped surface will have a larger impact on measurements made on smaller-scale maps. This is because, for the same millimetre square of map sheet that is stretched or contracted, on a smaller-scale map this millimetre corresponds to a larger area on the ground.

Most of the data that are incorporated into a GIS were initially in analogue form and, therefore, had to be digitised in order to be used in the GIS. This will introduce further errors into the map data (Bolstad et al., 1990). In order to calculate the overall error of the data held in a GIS, the errors associated with the several production stages of the map must be combined with the error introduced by the analogue to digital conversion. To assess the total error associated with producing a map is, however, 'a very difficult, if not impossible, task' (Thapa and Bossler, 1992, p. 839), because the functional relationships among the various errors introduced at different stages of the mapping processes, dimensional instability of the medium, and the original digitising procedure are often unknown. The explanation of how different individual errors can be combined to provide a total error value, together with the estimation of the non-generalisation error of the maps used in this study, is given in section 4.1. As the main objective of this book is to measure the generalisation effects, it is crucial that other sources of error are taken into account in order to try to isolate the magnitude of the error due only to generalisation.

2.2 Generalisation effects

In describing generalisation effects, one of the most useful distinctions is between generalisation for analysis purposes and generalisation for display purposes – termed, by Brassel and Weibel (1988), statistical and cartographic generalisation. The term 'statistical generalisation' has led at times to misconceptions and, recently, the equivalent term *model generalisation* has been introduced instead (Weibel, 1992). Generalisation tools for analysis and display purposes should both be included in a GIS.

Model generalisation is mainly a filtering process. According to Brassel and Weibel (1988) it is never used for display, but strictly for data reduction in order, for example, to obtain a subset of an original database for data analysis. This type of generalisation has no equivalent in manual generalisation and is therefore unique to digital systems. In contrast to this, *cartographic generalisation* is used for graphic display, and it aims to improve the visual effectiveness and readability of a map (Brassel and Weibel, 1988). This type of generalisation is subject to the same principles as those which apply in manual generalisation (Weibel, 1986).

The relationship between cartographic and model generalisation is illustrated in Figure 2.2. In order to clarify the distinction between the two types of generalisation, a further set of terms has recently been introduced, mainly by European researchers (e.g. Müller, 1991b; Grünreich et al., 1992; Weibel, 1992). These terms are necessary because the use of GIS has widened the number of possible permutations of databases. A Digital Landscape Model (DLM) is a generic term for a comprehensive description of the landscape, usually in the form of a topographic basic scale map. This primary model of the real world is based on the acquisition of original information by a topographer or a photogrammetrist. Even at this basic level, the modelling of the real world can be considered to involve a certain degree of generalisation, in the form of representation, selection and abstraction, and has therefore been called 'object generalisation'. Databases derived from the primary DLM, through *model* generalisation, are special-purpose secondary models of reality, such as thematic maps that describe different spatially distributed phenomena. Secondary DLM are likely to be used solely for analysis and they are not intended for graphical representation. Therefore, scale is not as important and

accuracy becomes a better criterion. The accuracy parameter that character-
ises the primary and secondary models (the parameter A_x in Figure 2.2) is
related to the data source, and indicates either the level of detail or the
maximum error of the database (Brassel and Weibel, 1988).

Both primary and secondary DLM may be used to create a cartographic
representation, or a Digital Cartographic Model (DCM), through the process
of cartographic generalisation (Müller, 1991b). The drive towards a scale-
independent GIS would mean that there would be an ability to perform an
instantaneous update of DCM from a central database (i.e. a DLM). A DCM
can also be derived from other DCM by cartographic generalisation. This is a
common procedure used in mapping agencies when, for example, a 1 : 200 000

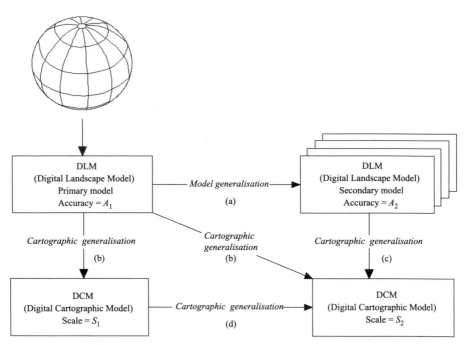

Figure 2.2   The relationship between Digital Landscape Models (characterised by a specific
thematic content and a related accuracy parameter A_x) and Digital Cartographic Models
(characterised by a specific scale S_x): (a) a DLM can be generalised by reducing it into a DLM
of lesser information content (e.g. $A_1 = 10$ m resolution and $A_2 = 50$ m resolution) – this
conversion is called *model generalisation*; (b) a DLM can be generalised by converting it into
a different DCM which may be displayed at different scales (e.g. $S_1 = 1 : 50\,000$ and $S_2 =$
1 : 250 000) – this conversion is called *cartographic generalisation*; (c) a DCM can also be
derived by cartographic generalisation from a secondary DLM; (d) finally, a DCM can be
derived from other DCM by cartographic generalisation (after Brassel and Weibel, 1988;
Müller, 1991b).

scale map is derived from a 1 : 50 000 scale map which, in turn, is the result of generalising a 1 : 25 000 scale map. Different DCM are characterised by the different scales at which they are intended to be displayed (the parameter S_x in Figure 2.2). In a GIS, both DLM and DCM are spatial databases.

There is another conversion that is possible in theory, but in most circumstances at present would not be legitimate. This would be to convert from a DCM back to a DLM. This would entail the 'reverse engineering' of the effects of cartographic generalisation, because otherwise there would be a reduction of accuracy, and in most cases this would not be acceptable. As Vitek *et al.* (1984) point out, the process of generalisation introduces error by eliminating detail for portrayal at a smaller scale, but this detail can not be recreated simply by enlarging the maps. The difficulty in 'reverse engineering' generalisation means that the conversion from a DCM to a DLM is absent from Figure 2.2.

GIS need to perform cartographic and model generalisation in order to satisfy both display and analysis purposes respectively. For the display subsystem, emphasis should be put on optimal graphic communication and on conventional cartographic generalisation; while 'in the analysis and modelling subsystem, attention should be paid to optimal preservation of data characteristics, and the data should be changed as little as possible' (Weibel, 1986, p. 22). Since model and cartographic generalisation have such very different purposes, they cause very different generalisation effects in the data, and they both affect the accuracy of spatial databases in their own way (Müller, 1991b).

Model generalisation should cause fewer generalisation effects because the data reduction is done under quantitative-statistical control. The reduction of data volume is maximised while at the same time modification of the source data is minimised; for instance, derivatives (e.g. gradients) need to be preserved at sufficient accuracy (Weibel, 1992). Because of this, model generalisation relies exclusively on selection and simplification generalisation operations and excludes displacement – in other words, generalisation becomes a simple act of *smoothing* (Weibel, 1986). Model generalisation aims therefore at minimum average displacement, and concepts of reliability, tolerance or error may be applied (Brassel and Weibel, 1988).

On the other hand, cartographic generalisation will cause more generalisation effects than model generalisation, because it operates by employing *ad hoc* decisions involving bias and elimination, enhancement and shape distortion, and is therefore non-statistical (Brassel and Weibel, 1988). Cartographic generalisation not only involves the generalisation operations of selection, simplification and smoothing, but it also includes feature displacement (Weibel, 1986). Its purpose is not minimum average displacement, but representativeness in a graphical rather than an analytical sense: 'It might cause displacements, distortions, and exaggerations of map elements locally, if that is needed to preserve the characteristic look of the map' (Weibel, 1992, p. 134). Because of these different generalisation effects caused by model and cartographic generalisation, it is important in a GIS always to specify the purpose and the need for generalisation.

Model generalisation is especially important in the context of GIS: 'Possible applications include eliminating redundant data elements, controlling data reduction to speed up subsequent model computations, and adjusting the accuracy of different data sets that are to be processed jointly (e.g. in polygon or data merging)' (Weibel, 1992, p. 134). The primary DLM often needs to be compressed and its content reduced in order to run the analysis faster (but with as little information loss as possible), or in order to save storage capacity (Weibel, 1986). In such a context it is suitable to use only elimination and simplification operations guided by statistical criteria (rather than visual effectiveness), and to change the remaining data as little as possible in the generalisation step.

Model generalisation might also be needed in order to increase the accuracy of the data: this is generalisation as an error reduction mechanism. As discussed in the previous section, errors in spatial databases occur at all stages, from data collection to data manipulation. Increased precision in measurement or more extensive data sampling might increase the chances of error in interpretation, because the true value of a single observation, the measurement of which is affected by random errors, may be hidden through some high-frequency disturbance (Müller, 1991b). Painho (1995), for example, found that class aggregation generally had the effect of increasing the attribute accuracy of natural resource maps. Jean-Claude Müller added that generalisation is needed 'in order to filter out the errors and consolidate the trends. A generalised trend is more robust than an individual observation. Smoothing operations to generalise curves and surfaces make this assumption; for instance, there is the notion that if observations had been more accurate the curve would have been smooth' (Müller, 1991b, p. 458).

Another situation in which generalisation can reduce errors is in the case of the creation of sliver polygons, which, as previously mentioned, is a generalisation-related problem specific to vector-based systems. Michael Goodchild described the creation of sliver polygons when overlaying five coverages in the Canada GIS (CGIS). According to the author, the number of polygons produced in the overlay depended on the complexity of each boundary and not so much on the number of polygons overlaid. Goodchild added that, paradoxically, 'generalisation of boundary lines tends to reduce the spurious polygon problem rather than increase it' (Goodchild, 1980b, p. 202). More generalised features will generally have less detail and therefore, when overlaid, will generate a smaller number of sliver polygons.

However, generalisation done for analysis purposes can create undesirable generalisation effects: 'A wrong classification, for instance, may hide the characteristic pattern of a statistical surface. Furthermore, a classification may create crisp boundaries between areas whose boundaries are naturally fuzzy' (Müller, 1991b, p. 460). Model generalisation, especially the grouping of data by larger areas, is also known to affect inter-variable comparisons. Openshaw (1984) demonstrated this for socio-economic data, as did Burrough (1983) for the spatial variation of soil data. This concerns generalisation problems that are of a statistical (rather than geometrical) nature, such as the Modifiable Areal Unit Problem (or MAUP). The concept of the MAUP dates at least as far back as the 1930s (see, for instance, Gehlke and Biehl, 1934; Neprash, 1934) but has become recognised to be of central importance in GIS-related work (Openshaw, 1984). Two different cases are involved: the effects of zonal 'reporting units' at different levels of spatial aggregation (such as tracts, counties and states) and the effects of different arrangements of boundaries for a single level of aggregation (leading to the gerrymandering concept).

Openshaw (1984) showed that spatial relationships, established through simple techniques such as correlation between zone values for different variables, may lead to wildly different answers and hence interpretations. This is not solely a matter of aggregate statistical reporting; the spatial patterns resulting from different aggregations may differ greatly, whether the assessment tool is statistical or the human eye. It should be noted, however, that Tobler (1989) has argued that the problems arise chiefly from using inappropriate methods: he argues that MAUP could be overcome by the use of 'frame invariant' approaches. On the other hand, Openshaw (1996) suggests that, rather than viewing MAUP as a nuisance, which is best removed or controlled in some way, MAUP can be an invaluable tool. He proposes the use of zone design as the basis for a novel approach to spatial analysis and the modelling of data. As a pattern detection and modelling device, it can both visualise complex spatial patterns and model spatial relationships.

Although errors can be introduced by model generalisation, these errors will always be inferior to those introduced by cartographic generalisation due to their distinct objectives. Since they are a result of cartographic generalisation, DCM are not as reliable as DLM (Müller, 1991b). Because model and cartographic generalisation generate different levels of accuracy, the data resulting from them should be used in different ways: 'One cannot expect a database which was generalised using procedures for cartographic generalisation to be a reliable source of predictable quality for analysis, because manipulation by cartographic generalisation will introduce unpredictable errors through processes such as feature displacement' (Brassel and Weibel, 1988, p. 236). However, quite often, spatial databases are DCM rather than DLM, since they were digitised from existing maps: 'The current strategy is to produce DCM which are an exact replica of the topographic map series. The resulting products are scale dependent and "tiled" to the traditional map sheet format' (Müller, 1991a, p. 258).

Current GIS spatial databases are often created from digitising existing analogue maps which, in turn, have been made by manual cartographic generalisation methods. As a consequence of this, most digital databases have generalisation effects embedded within them that have been derived from

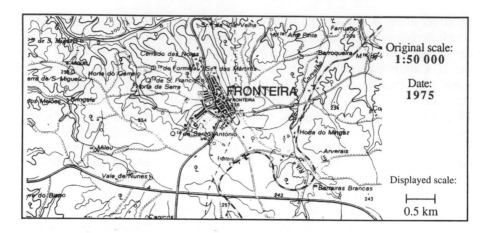

Original scale:
1:50 000

Date:
1975

Displayed scale:

├───────┤
0.5 km

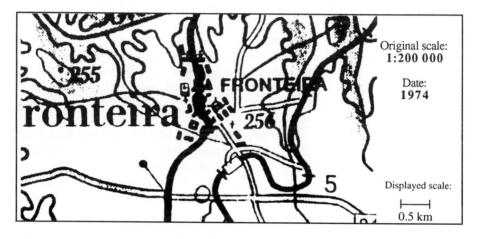

Original scale:
1:200 000

Date:
1974

Displayed scale:

├───────┤
0.5 km

Original scale:
1:500 000

Date:
1981

Displayed scale:

├───────┤
0.5 km

traditional manual generalisation. These generalisation errors (locked into analogue maps and subsequently propagated into digital data sets) are the subject of this book. It is clearer to illustrate some of the effects of manual cartographic generalisation processes with a specific example, taken from the topographic maps of one of the study areas of this research. This will be descriptive rather than quantitative.

In Figure 2.3 is portrayed a portion of a 1 : 50 000 scale map which covers the area around the town of Fronteira in the Portuguese district of Portalegre. The maps at the scales of 1 : 200 000 and 1 : 500 000 are generalised versions of the 1 : 50 000 topographic map, and are enlarged to match the 1 : 50 000 scale map for comparison. All of the maps are therefore displayed at the same scale, but the scale from which the three maps are derived is progressively reduced. It is important to note that, although the following comparisons of the smallest-scale maps are made in relation to the 1 : 50 000 scale topographic map, this map is also a generalised version of reality.

The representation of Fronteira, a town with 2606 inhabitants (1981 population census), is progressively generalised until, at the smallest scale, its location is symbolised by a circle. Several changes in the representation of the town can immediately be observed. In general, the numbers of buildings or groups of buildings shown are reduced. Groups and blocks of houses are progressively regrouped (except in some cases where there is a road between them) and isolated houses are eliminated or enlarged. Certain important landmarks retained at the smaller scales are symbolised and enlarged (e.g. the cemetery – represented by a square with a cross, in the north-west part of Fronteira – increases in size from the 1 : 50 000 to the 1 : 200 000 scale maps).

Due to constraints of human drawing and perception, there is a reduction in the amount of detail depicted from larger scales to smaller scales. According to Müller (1990a), generalisation also causes a change in the ratio between open rural space and the areas occupied by man-made features (due to the exaggeration of the graphic representation of roads and built-up areas). In the case of Fronteira, the town symbol actually takes up less space at the 1 : 500 000 scale, but the massively widened roads mean that, overall, human space takes over more of the map.

Müller also argues that the topological relations are better preserved than geometric aspects of the map (e.g. length or form). While the geometric properties of the Fronteira map have certainly been changed, there are also some serious topological differences. On the 1 : 500 000 scale map, the road 243

Figure 2.3   The area around the town of Fronteira, Portugal, as portrayed on topographic maps at the scale of 1 : 50 000 (top), 1 : 200 000 (middle) and 1 : 500 000 (bottom). The scales of 1 : 200 000 and 1 : 500 000 are enlarged to match the 1 : 50 000 scale map for comparison. (Source: Instituto Português de Cartografia e Cadastro.)

running east–west, unlike at the scale of 1 : 50 000, does not go straight across the main road running north–south, but has a small step in it. Moreover, it now seems to be mislabelled. The '243' label is closest to the road leading directly from Fronteira, while the 243 is actually the road running along the bottom of the map. Also, with decreasing map scale, Müller (1990a) argues that the *number of classes* of objects displayed on the map decreases while the *density* of objects per square map unit increases. The number of classes clearly decreases, but the density of objects follows no set pattern in the Portuguese area. In fact, the smallest-scale map has the lowest density of objects per square map unit.

The linear elements of the maps of Figure 2.3 also suffer different kinds of generalisation. The most important linear elements (e.g. major roads, railways and main rivers) are progressively widened and generalised; that is, are less 'wiggly'. Roads in particular are disproportionately exaggerated compared with other linear elements such as rivers. Contours are progressively removed and the remaining ones become more and more simplified; minor features (e.g. smaller roads and rivers) are eliminated. The reduction of features at the smaller scales demonstrates how feature selection helps the cartographer to avoid clutter.

The generalisation of lines coupled with the reduction of their sinuosity and length can be observed qualitatively in the transformations to the railway track south of Fronteira. The characteristic loop of this part of the railway is maintained by the cartographer at all scales, but its shape and size are altered due to constraints of human drawing and the reduced space in which to represent it. These changes of shape are of two types: straightening of certain parts of the loop, and an increase in the angularity of certain bends. At the scale of 1 : 500 000 there is a clear displacement of the gap of the loop to avoid coalescence of the sides of the loop.

Other authors have assessed the results of map generalisation visually by observing maps showing the same areas at different scales. Buttenfield *et al.* (1991), for example, proposed that, by comparing the representation of single features at multiple scales, a systematic inventory of existing map series would allow the identification of consistent treatment of cartographic objects. Such qualitative comparisons between topographic maps (e.g. Müller, 1990a; Buttenfield *et al.*, 1991) have been done with the main motivation of formulating generalisation rules, rather than the analysis of generalisation effects *per se*. And while they have proposed qualitative generalisation rules (such as 'the density of objects increases with decreasing map scale' – Müller's rule mentioned above) exceptions often mean that the rules are not universally applicable. As a humorous aside, Müller (1990a, p. 320) proposed the rule that 'all settlements under 3000 people may not be represented at a scale of 1 : 1 million or smaller', because his own study town disappears at the scale of 1 : 1 million and, as a consequence, he could not take his analysis any further. However, the map of Iberia in *The Times Concise Atlas of the World*

(Times Books, 1990) at the scale of 1 : 3 million, still shows Fronteira (with a population under 3000). This highlights the argument presented in the previous chapter, and continued in later chapters, about the difficulty of designing universal generalisation rules.

These qualitative studies have only compared topographic maps visually. However, it is important to proceed to *quantitative* measurements which give a more precise picture of the impact of generalisation effects on data quality. In the following section, previous research into generalisation effects is divided into four main categories. The first is the completeness of a database. A second type concerns the differences in measured lengths, areas and surfaces. The third category of generalisation effects is the shift or displacement of the map features, which affects positional accuracy. Finally, past studies on the impact of generalisation on GIS-based analyses are reviewed.

2.2.1 Generalisation and the completeness of a map

Competition for map space is a fundamental principle of map design. Unwanted features are eliminated and important features are retained. The elimination of features is an extreme, though common, case of generalisation. It is one of the most immediate differences between maps of the same region at different scales, as can be seen in Figure 2.3. It follows that the completeness of a map is affected by the elimination of features due to generalisation procedures.

In traditional cartography, various steps and *ad hoc* rules are used to decide which features are selected at different scales. For example, when generalising a network of drainage channels the main channels will have priority. If there is space, smaller streams are chosen next, depending on their length or importance (e.g. short interconnections between major drainage channels might be retained; Keates, 1989). However, there have been attempts to make the steps of selective omission more consistent. One of the key realisations was that the amount of information that can be shown per unit area of any map decreases according to a *geometrical* progression.

This was formalised in the 1960s in an attempt to quantify the elimination of features from maps at different scales. Töpfer and Pillewizer (1966) measured the difference in the number of features (e.g. islands off the coast of Yugoslavia) found on a series of small-scale maps of different countries in order to determine the number of objects that should be selected for depiction at a smaller scale. The motivation for these measurements was to establish a rule that future cartographers could follow, which would help them to decide how many features (e.g. number of railways) they should show at different scales. On the basis of these measurements, the authors went on to determine whether other maps conformed to Töpfer's *Principle of Selection* and which constituted exceptions to the rule. The Principle of Selection (also known as the Radical Law of Cartographic Generalisation), first proposed by Töpfer in 1961 (Töpfer and Pillewizer, 1966), is expressed by an equation that relates the number of occurrences of a particular feature at a source map scale and at a derived map scale. The principle can be expressed in its simplest form as (Töpfer and Pillewizer, 1966):

$$n_f = n_a \sqrt{M_a/M_f} \qquad\qquad (2.1)$$

where n_f is the number of objects that can be shown at the derived scale, n_a is the number of objects shown on the source material, and M_a and M_f are the scale denominators of the source and the derived map respectively.

Töpfer and Pillewizer argued that this equation worked best at medium scales up to 1 : 1 million. However, when symbols become more abstract in small-scale atlas cartography and conform less to the dimensions of the objects that they depict, two additional constants had to be introduced (a constant of symbolic exaggeration and a constant of symbolic form). The various case studies showed that it was comparatively easy to provide a measure for quantifying the information that a cartographer could reasonably expect to put on a derived map.

What the law failed to address is *which* particular settlement or railway should be selected. Mark (1990) used Töpfer's constants as a basis for 'importance values' in an attempt to get around this limitation; in other words, to decide which features should be selected within an automated generalisation system. Töpfer's law could then be brought in to check whether enough features had been eliminated. Despite being 30 years old, Töpfer's law has been the *only* quantitative rule that has looked at the selection and elimination of features. In this study the results are analysed in the light of Töpfer's law (see section 5.3). Most other quantitative generalisation research has concentrated on the effect of generalisation on measurements of length and area.

2.2.2   Generalisation and measured lengths, areas and surfaces

It has been known for a long time that the same line features on maps at different scales will have varying lengths. Penck (1894) showed that the length of the coastline of Istria varied from 223 km on the 1 : 75 000 scale map to only 105 km for the 1 : 15 million scale map. He argued that it was the removal of many of the coastal indentations that was apparently shortening the coastline at smaller scales (as described in Maling, 1989). Maling (1989) carried out further work in the late 1960s and showed large differences (although not as extreme as Penck's) for the coast of Yorkshire. Various other researchers have measured line lengths around the world – the coast of the Black Sea (Vorob'ev, 1959), the shoreline of Scandinavian lakes (Håkonson, 1978), the coastline of Australia (Galloway and Bahr, 1979) and many others.

Penck's (1894) solution to the problem of variable length was to make all the measurements at the same scale, so that the distortion was kept constant. However, this is not necessarily straightforward if different authors or cartographers were involved. Also, it might not always be an acceptable solution, if a uniform coverage is unavailable or is only available for very small scale maps which may have been excessively generalised (Maling, 1989).

The corollary of decreasing length of a line with decreasing map scale is variability introduced by different measurement techniques for the same map. Richardson (1961), in his study on the causes of wars, noted that the recorded length of the common land frontier between Portugal and Spain was different in their respective encyclopaedias (1214 km versus 987 km). Richardson put this down to differences in the divider's opening size used to measure the frontier: the smaller the gap between the dividers, the longer the frontier becomes. He went on to measure the Australian and South African coasts, the west coast of Britain and the German land frontier. The results of this sort of analysis are very similar to the findings from generalisation due to different map scales. For instance, the more wiggly and complex the line was, the greater was the effect. Richardson mathematically related the results to the equation $L \propto D^{-\alpha}$, where L is the total length of the line, D is the width of the dividers in map units, and α is a positive constant characteristic of the frontier. This is sometimes known as the 'Steinhaus paradox', after Steinhaus (1954) who investigated it mathematically.

All of the empirical studies so far have tended to concentrate on one particular feature of the map; for example, coastlines, or shorelines of lakes. And while there is variation within features (cf. the coast of South Africa with the west coast of Britain), there will often be even more variation between features. Different line features (roads, railways and rivers) of a particular area will have their lengths changed in different ways across different scales. One of the aims of this book is to investigate this phenomenon (see section 5.1).

The difficulty in pinning down a particular length for a line, both philosophically and empirically, led to the involvement of fractal analysis. Pioneered by Mandelbrot (1967, 1982), fractals within geography are a continuation of Richardson's work. Mandelbrot redefined α in the above equation as the fractal dimension, D. (Mandelbrot's use of D is not the same as that of Richardson.) For any line, D can fall between 1 and 2 inclusively: $D = 1$ for a completely smooth line and $D = 2$ for a line that is so complex that it covers an area (the geographical equivalent would be the shoreline of a lake that is so convoluted that it fills the lake!). Virtually all geographical lines fall in between these two integers, and hence are *fractal* dimensions. The importance of this concept was that it gave a measure of a line that was apparently independent of scale. Features that have the same fractal dimension across different scales are termed self-similar. In other words, the gross pattern of a line at smaller scales is replicated at finer scales. Features that are not self-similar are termed scale-dependent.

There is some disagreement as to the importance of fractals when measuring lines. Goodchild and Mark (1987, p. 267) say that D 'may be the most important single parameter of an irregular cartographic feature'. In contrast, Krantz sarcastically states that 'one notable difference between fractal geometry and calculus is that fractal geometry has not solved any problems' (Krantz, 1989, p. 14). He goes on to warn about researchers jumping on the 'fractal band wagon' without enough forethought about what they are really trying to measure.

While self-similarity is an intuitively attractive parameter for describing the characteristics of spatial data, some recent work has indicated that fractal-derived surfaces do not have many of the characteristics of those found in, for example, geomorphology. Moreover, irregular surfaces need to be estimated statistically, and Batty et al. (1989) have shown that better methods for doing this are required. Considerable uncertainty still exists, therefore, about the merit of fractals in studies of generalisation. More importantly, it has now been realised that D for geographical features is often not constant across scales, as was originally expected. Goodchild (1980a) found that the east coast of Britain had a low D (i.e. smooth lines) at large scales but a relatively higher D (i.e. jagged lines) as the scale decreased. Müller (1986) studied seven coastlines, two lake shores and one river in North America across different scales. He found that overall (in contrast to Goodchild) the fractal dimension decreased as the scale became smaller; in other words, the lines were smoothed out. Müller argued that this was a result of generalisation, and that the preservation of fractal dimension should be used as a standard for generalisation algorithms.

This standard rests on the assumption that the fractal dimension does remain constant in the real world. However, is there any evidence for this assumption? A counter-explanation would be that the decrease in the value of D in Müller's work is an accurate reflection of what happens in the real world. Mark goes further and argues:

> I do not think that it is reasonable to claim or believe, that there are actually fractals in the natural 'real world'! Fractals are no more 'in' the real world than normal curves or Euclidean geometry . . . . Thus, while scale-dependent geometry is important to GIS, I don't think that fractals *per se* must be taken into account at a fundamental level. In particular, I don't think that the existence of fractal mathematics detracts in any way from classification of geographic phenomena into points, lines, regions, surfaces, and solids, based on integral dimensions (David Mark, personal communication, E-mail discussion on fractals transmitted on 8 September 1990).

This book concentrates on the more concrete aspects of line length, and only occasionally are fractal dimensions used as an additional measure of feature characteristics.

Just as the length of lines are changed by generalisation, so areas are also affected across different map scales. For example, boundary lines of a particular zone will usually become less wiggly as the map scale decreases, influencing the total area. But generalisation can also affect areas in another way. At different scales the area of relative features will be changed. Müller (1989) noted that as the map scale decreases the map area devoted to built-up space relative to rural area increases; in other words, man-made objects take up more of the map space.

Raper et al. (1992) quoted a study in which the area of Bristol that was built-up was estimated. In the study it was recognised that the size of this area was highly dependent on the scale chosen: 'As the scale of the map increases say from a 1 : 250 000 road atlas to a 1 : 10 000 scale plan it becomes possible to separate parks and major open space from the surrounding areas, and with further increases in scale – up to the 1 : 1250 scale – to separate houses from gardens' (Raper et al., 1992, p. 221). To overcome this problem, a completely different method was chosen, in which postcodes were used to estimate the built-up area. By using all of Bristol's postcode zones (of 100 m resolution) and by defining built-up areas as having six or more delivery points, the study was able to plot built-up locations and calculate the total area. It was also possible simply to vary the number of delivery points (e.g. ten or more) to explore how the cut-off point affected the total area. According to the study, this estimation of area for the built-up zones of Bristol was more flexible and accurate than for an area measured more conventionally at any one particular scale.

An indirect consequence of the simplification of line detail is the alteration of shapes. Conversely, shape change, rather than length or area change, can indicate that generalisation of features might have taken place. Leitner and Buttenfield (1995) measured the amount of simplification present in the Austrian national topographic map series by counting the number of settlement objects (e.g. houses) that had been simplified from an irregular to a regular shape. This was carried out to extract generalisation rules. For example, from a 1 : 50 000 map sheet they formulated the following rule: 'If scale changes from 1 : 50 000 to 1 : 200 000, and the block-type is built-up, then four times as many residential buildings are simplified at 1 : 200 000 than at 1 : 50 000' (Leitner and Buttenfield, 1995, p. 239).

Generalisation also has an impact on the measurement of surfaces; for example, the terrain profile. Weibel (1992) has carried out some initial investigations into the effects of generalisation on digital terrain models of Swiss mountains. He found that different terrain types were altered in different ways. For example, simplification tended to be greatest along ridges and channels; that is, they were smoothed out. While Weibel carried out some quantitative assessment, in his study he used mainly visual evaluation, arguing that, although subjective, it was a valid part of assessing *cartographic* generalisation.

Although there has been some work on the effects of generalisation on areas and surfaces, the vast bulk of the research has been carried out on lines.

This is reflected in this book in that most of the features investigated are line features, as there was less scope for assessing the effects of generalisation on areas and surfaces (the only well defined area features were the urban zones).

2.2.3   Generalisation and positional accuracy

Another important aspect of the effects of generalisation is the shift or displacement of the map features, which in turn affects positional accuracy. Positional accuracy is a measure of how closely the position of map features reflects the true position of the features on the ground. Much effort is expended in ensuring that a few key points (such as triangulation points) are as accurately placed as possible across all map scales. However, many other features may be shifted from their true position, and this shift will generally grow as the scale of the map decreases.

An example of a standard for map accuracy is the one used by the United States Geological Survey. A typical rule is that for maps of 1 : 24 000, 90 per cent of the points tested should lie within 40 feet of the true position on the ground (from Maling, 1989). The fact that points are used as a standard, rather than other features such as lines, simplifies the procedure for testing accuracy, although these other features are just as likely to lose positional accuracy (Chrisman, 1991).

Positional accuracy is affected by cartographic generalisation because features may be displaced and their original shape may be distorted. In maps with a greater degree of generalisation, 'the relatively few selected content elements will be represented by strong expressive forms, which can be relatively far away from their original positions' (Swiss Society of Cartography, 1987, p. 17). Reduction of the positional accuracy can in turn affect attribute accuracy: 'As the locations of objects become more uncertain, their attribute characterisation becomes fuzzier and more complex' (Müller, 1991b, p. 459).

Besides affecting positional and attribute accuracy, shifts of geographical features between different scales can cause further problems in a GIS. The location of geometric centroids can move, which in turn can alter interzonal distance calculations. Also, even within a single map coverage the ratio of areal distribution (e.g. the percentage of an area occupied by different soil types) can change if the lines between the different zones are shifted. Therefore, at smaller scales, in addition to the positional error of the boundaries, there can be an attribute error and some feature classes might be wrongly labelled. Beard (1988) found that when evaluating the manual generalisation of the coastlines of Maine, North Carolina and South Carolina, 2 per cent, 7 per cent and 36 per cent of *water*, respectively, was classified as *land* area on the 1 : 250 000 scale generalised map.

The way in which generalisation affects the positional accuracy of the different geographical features varies according to the feature type and the density of features within the map space. Some features can actually change in position over time. For example, a road has a semi-permanent existence; in contrast to a river, that has a position that can fluctuate depending on the season. The positions of some artificially imposed features, such as political borders, can also be vague (Fisher, 1989). Also, many geographical phenomena, such as soils, have 'fuzzy' boundaries, so although they may be represented by precise lines on maps, in the real world the boundaries are not so clearly defined.

Deliberate feature displacement normally occurs if two or more features to be portrayed on a map are so close together that when they are displayed in their proper position they overlap. According to Thapa and Bossler (1992), in actual ground terms, quite large shifts can occur due to generalisation depending on the map scale and the proximity of the features to be portrayed (although they did not give figures). The Australian Survey proposes that on their 1 : 250 000 scale topographic database, when two features are almost coincident and one must be moved to ensure clarity, a typical displacement of roads and railways could be up to 200 m: 'In the worst case, where five of these features are close and adjacent, one may have been displaced by up to 675 metres' (Australian Survey, 1992, p. 86).

Features that are present in the denser areas of a map will be affected by the displacement to a greater extent than features that are in less dense areas of the same map. Also, even in the dense zone of a map, different feature types will be affected differently. National mapping agencies usually have an established hierarchy by which features should be displaced. For example, in the case of conflict between features, the rules for the Dutch mapping agency indicate that roads should be displaced first. If a conflict still persists, then rivers should be displaced, and only as a last resort will railways be displaced (Müller and Mouwes, 1990). In contrast to this, the Australian mapping agency displaces rivers first (Australian Survey, 1992). Therefore, different feature types are affected differently according to which mapping agency produced the data.

The shift of features has a larger impact on the use of a GIS when multiple coverages are combined. The well-known problem of 'sliver polygons' is a manifestation of this problem (Goodchild, 1980b). Rhind (1988), for instance, described the difficulty in making the 1 : 63 360 scale geology digital map of Merthyr Tydfil in 1973–75. The geological boundaries were surveyed at 1 : 10 000 scale and were then overlaid at the smaller scale to fit over a manually generalised topographic base map. However, due to generalisation, there were gross discrepancies between the OS 1 : 10 000 and 1 : 63 360 scale maps. Not only were statistical errors generated in the resulting two-dimensional coverage, but the topological errors generated inadvertently led to a locally different three-dimensional geological model: 'The critical relationships between topography and geology, which enable the map user to understand the three dimensional nature of the substract, were periodically and unpredictably violated. All of this was a consequence of the "errors" in the smaller scale and manually compiled topographic base map' (Rhind, 1988, p. 284). An example of this was the situation in which outcrop boundaries in 1 : 10 000 scale mapping fitted within the 'walls' of a railway cutting at 1 : 63 360 scale. The width of the outcrop was scaled correctly, but the equivalent cutting shown on the topographic base map was both relatively wider and differentially displaced, usually on one side of rather than equally about a median line. Attempts to rectify these sorts of discrepancies, that were more than just sliver polygons, probably trebled the cost of the Merthyr Tydfil project (João et al., 1990). While discrepancies over scales such as these have often been noted, there has been little empirical research into the effect of generalisation on positional accuracy.

2.2.4   Generalisation and GIS data manipulations

The previous sections have looked at the effect of generalisation on particular measurements. However, some research has been directed at the effect that different source scales will have on the result of GIS *analyses* or *map manipulations*. While most GIS researchers are aware of the potential for generalisation to introduce error, they often merely acknowledge the problem and then fail to take it into account in the final analysis. However, a few researchers have studied this in its own right.

SERRL (South East Regional Research Laboratory; see Masser, 1990) carried out such a project. They compared British urban area boundaries at the 1 : 10 000 and 1 : 50 000 scales. In order to determine whether the finer scale made a difference to analyses using the urban boundary data, they carried out a series of point-in-polygon operations. The polygons were the district boundaries and the points were the centroids of the enumeration districts of the UK 1981 population census. The estimated populations in each district were compared for the two scales. The study revealed that there were differences in the calculated populations between districts due to the generalisation of the district boundaries and the position of the points (only accurate to within 100 m).

Part of the goal of the project was to determine whether the extra cost of digitising the 1 : 10 000 maps was worthwhile in terms of increased accuracy (i.e. the differences between the scales were large enough to be concerned about). SERRL emphasised to their client, the Department of the Environment (DoE), that this was largely dependent on how they expected to use the data, and in the end the DoE decided to go ahead with digitising the 1 : 10 000 maps. In other words, the effects of generalisation on spatial analysis can have very real costs (John Shepherd, 1993, personal communication).

Investigating differences across scales is only one way of measuring the effects of generalisation. Blakemore (1983) took a different approach. He carried out a point-in-polygon analysis on data from the North West of England. The points were 23 000 industrial establishments and the polygons were 100 employment office areas. By introducing a zone of uncertainty (a Perkal band) around the lines defining the polygon boundaries, Blakemore re-created the effects of generalisation. Perkal (1958) defined an 'epsilon' distance (now also called a Perkal band) about a cartographic line as a means of generalising a line objectively. Blakemore (1983) suggested that this Perkal band could be used in reverse to indicate the width of an error band about a digitised polygon boundary. When the zone of uncertainty was set at 0.7 km, Blakemore found that there was considerable ambiguity in assigning each point to a particular polygon. Only 55 per cent of the 23 000 points were uniquely assigned, and there was doubt over the remaining 45 per cent of the points. Small and elongated polygons were the most severely affected. The author concluded that more attention should be given to the substantial errors that may result due to the generalisation process, which can have a considerable impact on the validity of geometric operations such as point-in-polygon searches. He felt that too much emphasis was placed on visual effectiveness rather than analysis when investigating generalisation effects.

The effect of generalisation on analysis within a raster format has also been investigated. Generally, researchers using a raster GIS are more aware of the consequences of their choice of cell size than vector GIS researchers are on the choice of the original source scale of their map data. There is more pressure on researchers using a raster GIS to justify why they have chosen a particular cell size and prove that it has not distorted their results. With research in the vector domain, there tends to be much less justification for a particular scale and the scale used is often the largest that is convenient. Blakemore (1985), for example, described the application of a vector map (originally produced for basic choropleth mapping) to tasks for which it was ill suited. The data were not accurate enough for detailed point-in-polygon operations, but were used anyway because they were easily available.

There are several studies using raster data that have proved that the increase of cell size (i.e. a form of generalisation) can reduce the accuracy of results. Wehde (1982) concluded that as cell size was allowed to increase, the accuracy of maps and inventories produced by computer processing decreased. Chang and Tsai (1989) studied the effect of Digital Elevation Models (DEM) resolution on spatial interpolation of slope and aspect. They concluded that DEM resolution functions in the same manner as map scale. High-resolution DEM had a higher mean (i.e. were steeper) and had a higher standard deviation (i.e. showed the bumps and dips better) for each slope, and resulted in more accurate slope and aspect maps.

Fellows and Ragan (1986) found that although the increase of cell size reduced accuracy, this reduction was dependent on the feature type and use of the data. The authors studied the interrelationships among cell size (e.g. of land cover and soils data), computer requirements and accuracy of results in hydrology-oriented GIS. The study concluded that the estimated volume of water run-off was relatively insensitive to increases in the cell size but, because of the role of slope in the timing of run-off, significant errors in peak discharge estimates could result as the cell size of digital terrain data was increased.

Goodchild (1980b) described some of the types of data errors that can occur in raster systems due to the effects of generalisation, such as in area and point estimation. According to the author, the standard error of both of these estimates is determined by the number of boundary cells of each patch and the cell size used: the larger the number of boundary cells and the larger the cell size, the more standard errors exist in area and point estimations. The number of boundary cells increases for boundary lines which are more 'contorted'; that is, have a higher fractal dimension D (Goodchild, 1980a). In addition, the number of boundary cells depends on the shape of the patch – 'a long, thin patch has more boundary cells than a circular patch of the same area' (Goodchild, 1980b, p. 195). According to Goodchild, the relationship between the fractal dimension of the boundary line and the increase in error with cell size is as follows: 'the more contorted the boundary, or the higher its dimensionality, the less rapid the increase in error with cell size' (Goodchild, 1980a, p. 92).

It is surprising that the work on the effect of generalisation with raster GIS has not prompted more concern about its role within vector GIS. Further research needs to be carried out to determine the consequences of generalisation on particular vector analyses. It is also important to determine how different generalisation procedures (both manual and automated) cause different generalisation effects on the data. The next section reviews the various studies carried out to investigate the effectiveness of algorithms that carry out generalisation.

2.3   Comparing generalisation effects of manual and algorithmic methods

The particular generalisation method used strongly influences the resulting generalisation effects. Manual generalisation causes different generalisation effects from the ones caused by automated algorithms. This was demonstrated by Beard (1988). The author found that detailed coastline data could be automatically generalised with less error than through manual generalisation. In addition, as McMaster (1987) demonstrated, different algorithms cause different effects on the data. However, the quality of each generalisation procedure cannot be judged in absolute terms, but only in relation to its actual use. This means that 'generalisation algorithms perform adequately if, and only if, they are used in the appropriate context (e.g. it makes a difference whether a line simplification algorithm is applied to a coastline or to a road). What is required is a control environment for the procedure which is based on an understanding of what should be achieved' (Brassel and Weibel, 1988, p. 232).

In any event, the results from one particular generalisation procedure are judged relative to other generalisation procedures. When new generalisation algorithms are first developed, their results are frequently compared with the results produced by existing generalisation algorithms, or by visual comparison with the results of manual cartography. However, with the exception of Beard (1988), very little work has been carried out on actually *quantifying* the differences between manual generalisation effects and the results of automated methods. Part of the research in this book covers this aspect, an important factor when it is considered that in the future more and more automated generalisation will replace manual practices.

Over the past 20 years, a variety of methods has been used to evaluate the effectiveness of different generalisation algorithms. However, most of the methods shared the criterion that it was important to retain at least some of the characteristics of a feature. For example, a river with large meanders should keep aspects of the meander shape at a smaller scale. Where the methods differed was in *how* that should be done. An early approach, for line simplification algorithms, was to evaluate whether the algorithms retained the important points for the line structure. Line simplification algorithms usually keep what are known as *critical points*, while redundant points are weeded out. Critical points 'include both those which are relevant to the physical characteristics of the line and those related to manufactured or "perceived" positions of importance, such as a city along a river' (McMaster, 1987, p. 331). Marino (1979) discovered that people (whether or not they were trained cartographers) were remarkably consistent in the selection of points that they considered to be perceptually important. She noted that these points were the ones of greatest angularity change. White (1985) compared lines produced by simplification algorithms with lines that had been simplified manually, in order to determine which algorithms retained more of these critical points. Of all the algorithms tested, the Douglas–Peucker algorithm was the most effective.

Later, researchers moved away from the principle of exclusively retaining critical points when trying to preserve the characteristics of a line. Thapa (1988), for example, argued that keeping all of the critical points resulted in simplified lines with many spikes. The problem of spikes in lines, which could actually lead to self-crossings (e.g. peninsulas that were closed off) was studied by Müller (1990b), who proposed a series of corrective actions that would eliminate the spikes after the algorithms were applied. Thapa (1988) concluded that, although the retention of some of the critical points was important, shape preservation should also be taken into account and, as a consequence, not all of the critical points should be retained. Therefore, the algorithm proposed by the author aimed at preserving the overall shape of the line rather than retaining critical points: 'Some of the critical points which are likely to cause spikes in the generalised lines must be eliminated if the generalised lines are to be smooth, uncluttered and aesthetically pleasing' (Thapa, 1988, p. 516).

The realisation that the retention of critical points was not enough to maintain the character of a line led some researchers to believe that fractals could provide the solution. Müller (1987), for example, argued that the preservation of fractal dimension should be the key criterion for the automated generalisation of lines. Müller observed that none of the algorithms published in the literature appeared to preserve with full certainty both self-similarity and fractal dimensions. Müller went on to propose a new algorithm based on the walking dividers method of calculating fractals.

Buttenfield (1989) also used fractals to analyse the difference between scale-dependent and self-similar linear features as the basis for determining the cartographic detail worth preserving during generalisation in order to maintain both geographical accuracy and 'recognisability'. Some authors have suggested that the fractal dimensionality should be pumped back into the line; that is, enhancing generalised lines so that they maintain the same signature as the original line. Dutton (1981) presented an algorithm that generalised the detail of digitised curves by altering their dimensionality in parametrically controlled, self-similar fashion. Dutton's algorithm did this by generating new points into a string of coordinate pairs, while maintaining the local self-similarity of the line.

However, Visvalingam and Whyatt (1990) were concerned that too much emphasis was being placed on the geometry of a line. They argued for a return to a more straightforward visual evaluation of a line: 'Does it look good?'. The Douglas–Peucker and other algorithms were analysed and the mathematical measures used to evaluate them were criticised. They concluded that objective statistical measures were not necessarily superior, and that much greater scope for concept refinement could be achieved through visualisation of the results of the simplification process.

Ehrliholzer (1995) went a step further and suggested that the ideal way to evaluate the quality of generalisation results is by integrating *quantitative methods* (such as measures of length) with a *qualitative evaluation* by cartographic experts (such as 'maintenance of the overall character of the original map'). The author argued that the use of a standardised questionnaire or checklist, that asks the evaluating expert to give grades to different assessment items, is the most effective way to handle the subjective aspects of generalisation. Ehrliholzer proposed combining the results of both the quantitative and qualitative assessment into a single quality score.

Concern with improving the graphic capabilities of generalisation algorithms has been reflected in the development of new algorithms that try to mimic the results produced by manual generalisation methods. For example, Li and Openshaw (1992) have proposed a new set of algorithms for line simplification which are better than previous counterparts at producing results equivalent to those produced by manual methods. The authors call this approach 'the natural principle of objective generalisation'. Other researchers (Müller and Wang, 1992) have also proposed new 'natural' algorithms for the generalisation of area-patches. The development of these new types of algorithms reflects a dissatisfaction with existing pseudo-generalisation algorithms for cartographic generalisation within the GIS community.

In addition, João et al. (1993) have emphasised the importance of the user being made aware of the impact of generalisation algorithms on the data and subsequent analyses. Generalisation effects, derived by using mathematical algorithms in an automated environment, are a function of: the format of the data (raster or vector); the generalisation algorithms chosen and the respective tolerances applied; the initial generalisation of the line (e.g. the number of initial points); the characteristics of the line (e.g. the 'wigglyness'); and 'how' the algorithm is applied (e.g. from which end of the line the algorithm starts). The number of algorithms chosen, and the combination and the sequence in which they are applied, will also affect the final generalisation effects.

Most current attempts at algorithmic generalisation have lacked the flexibility of a cartographer and will normally apply the same algorithm to the map as a whole without being sensitive to local variations of features. Evaluation of the structure and shape of features is important, as it can help determine where, when and why to generalise. In order to solve this problem, some designers of generalisation algorithms have recommended that the choice of generalisation procedures should be dependent on the geometry of the features. Buttenfield (1986), for example, suggested that to guide tolerance band algorithms for automated line simplification, such as the Douglas–Peucker algorithm, a procedure must be available that could adapt the tolerance criteria to the structure of the line to be processed. If the procedure were to run automatically, numerical descriptions of scale-dependent geometry and a consistent procedure for determining structural change at a variety of scales would be required.

More recently, Plazanet (1995) has described geometrical characteristics of linear features within a hierarchical classification. This was in order to obtain sections of lines that were geometrically similar enough to be handled by the same generalisation algorithm and parameter values. The hierarchical segmentation is based on the assumption that different levels of perception are applied to different shapes of a line; from the whole line, to an intermediate line section, and then down to an individual bend. Her method splits lines by determining key inflection points, and then classifies line sections according to their degree of sinuosity.

Automated generalisation can produce topological errors (such as the overlapping of features; Müller, 1990b) because algorithms such as the Douglas–Peucker algorithm only look at individual lines 'one at a time'. Manual generalisation typically considers the map as a whole, the cartographer moving one feature to give space to another. Automated generalisation methods are poor in the maintenance of relationships between features unless these are explicitly specified through a topological statement, typically polygon adjacency. For example, the Douglas–Peucker algorithm cannot cope with a situation in which a contour and a river to be generalised are required to remain in the same position relative to each other after generalisation (i.e. the river crossing the contour at exactly the same point). One solution might be to store the intersection points between features that cross each other. This was noted in an early reference on automated generalisation. Rhind's rule (Rhind, 1973) was to overlay all related features before automated generalisation, in order to maintain fixed points.

In automated cartography added human expertise is still necessary to overcome some of the problems caused by the inability of computers to handle features in the same way as people do. This is particularly important when dealing with relationships between different features within a given region. For example, a relatively flat and homogeneous area might well require a different generalisation procedure from that appropriate for a more mountainous and heterogeneous area. Weibel (1989) referred to this as a 'structure-dependent generalisation', in which the generalisation approach was adapted to (for example) different terrain characteristics. The underlying idea was that each geographical feature is significant in relation to its neighbours. Therefore, apart from measures of size, shape and orientation of isolated features, additional measures of interaction, density and spacing are essential for the selection of the optimum generalisation procedure (Weibel, 1986, p. 21):

> It is also necessary to conduct further research into the development of comprehensive methodologies for automated generalisation. Since all generalisation operations except selection result in some displacement of entire features or parts of features, it is important not to treat the individual feature classes (or layers) in isolation. If the layers are processed separately, it may well happen that undesirable interferences occur. For example, it is not acceptable to have a road placed in a lake.

Therefore, independently of how effective the algorithms turn out to be in dealing with individual features, it is also necessary to take into account the links between geographical features (e.g. rivers run perpendicularly to contours, and a bridge should be at the point at which a road crosses a river). To control the effects of generalisation properly, an holistic approach is needed. Rather than only treating digital features in a geometric sense (as with line simplification algorithms such as Douglas–Peucker), they should be seen as distinct geographical features within a particular context. The development of context-dependent automated generalisation is one of the key priority areas for generalisation research. The innovative work on the STRATÈGE platform by Ruas and Plazanet (1997) and on the MAGE system by Bundy et al. (1995) is helping to fill this research gap (see section 1.3.2).

In parallel with the above, it is crucial to develop methods to assess the quality of generalisation results. These are indispensable in order to evaluate the quality and performance of individual generalisation algorithms and whole automated generalisation systems. They are also important in the 'training' of automated generalisation systems based on machine-learning techniques (see section 1.3.3): 'The development of methods for evaluating generalisation results has received very little attention . . . it is only now being realised that such evaluation methods are an important component and even a prerequisite of knowledge acquisition' (Weibel, 1995b, p. 64). The OEEPE Working Group on Generalisation should play a key role in this respect. This group is currently coordinating an extensive study for the description and quality assessment of all currently available generalisation algorithms, including the development of quality criteria for a 'standardised evaluation of generalisation results'. It is planned that the evaluation of generalisation algorithms will also include 'undesirable changes'.

The consequences of generalisation are complex: the generalisation of one feature affects others and ultimately alters the quality of the end results derived from a GIS. The generalisation error is very difficult to quantify, because the amount of error introduced can depend on the type of feature and also on the character (or complexity) of the feature (Thapa and Bossler, 1992). Generalisation error can range from substantial for some features (e.g. coastlines) to non-existent for other features (e.g. completely straight roads). Painho (1995) found that it was impossible to speculate about the effect of generalisation on the accuracy of a map as a whole, as different classes responded in such different ways to generalisation.

Generalisation can potentially cause unintended transformations of the data that alter the topology of geographical phenomena, and affect subsequent statistical or geometrical calculations. Normally, GIS users would want to minimise, control and quantify the effects of generalisation on their results. Knowledge about the type and magnitude of generalisation effects embedded within spatial data sets should therefore be deemed essential by any GIS user. Despite this, almost all of the published literature only describes qualitatively the consequences of generalisation. Thapa and Bossler (1992, p. 838), for example, point out that 'substantial' shifts can occur in actual ground terms due to cartographic generalisation, but no figures are given. Other researchers (such as Beard, 1988) have measured generalisation effects of specific manually generalised features, but solely in order to evaluate automatic generalisation procedures. Blakemore (1983) and Goodchild (1980b) did investigate generalisation effects directly, but did not compare maps across different scales.

There are two critical research shortcomings at present. The first is empirical information on how serious generalisation effects are and under what circumstances they occur. The second is the need for better software – few GIS have mechanisms to cope with generalisation-related problems and none can produce a statistical summary of the changes to the overall characteristics of spatial data introduced by generalisation. The research challenge is therefore to develop an improved empirical understanding of the effects of generalisation via case studies. Any regularities that are found could then be used to provide a thorough description of generalisation effects. At present, the problems of generalisation are frequently discussed, but there are no authoritative estimates about their size and how best to measure them. The goal of this book is to determine the importance to GIS of generalisation-related problems, when and how they happen, and how best to handle them.

CHAPTER THREE

Selecting and preparing the maps

In this chapter, the study areas used, and the preparation of the maps that were necessary to define the magnitude and nature of generalisation effects in different types of geographical data, are described. A selection of map sheets at different scales from two different countries was used: an area around the town of Fronteira in the region of Sousel, in the centre of Portugal; and an area around the city of Canterbury in East Kent, in the South-East of England. Both regions had a mixture of rural and urban features. Two countries were chosen to highlight any differences in generalisation effects between the countries. Conversely, similarities would suggest the setting up of generalisation rules.

This chapter begins by providing a detailed description of the origin of the maps. Key to this was the lineage of the maps, such as whether the smaller-scale maps were derived from the larger-scale maps. A description of how the paper maps were digitised then follows. Because it was necessary for the measurements to be done using a GIS (see chapter 4), all of the maps had to be converted into a digital form. An interactive line-following digitising method ensured that the resulting digital maps were the closest possible representation of the original analogue maps. Finally, an explanation is given of the map transformations, the correction of systematic error, and the final edits that were performed. The emphasis is very much on a scientific reproducible method.

3.1 The maps used in this study: their selection and lineage

The maps chosen were topographic and were produced by the national mapping agencies of each country (for sample monochrome copies of the maps used, see Appendix C). The Ordnance Survey (OS) created the British maps and the Instituto Português de Cartografia e Cadastro, (the IPCC) produced the Portuguese maps. Topographic rather than thematic maps were chosen because they usually form the framework within which other field data are collected and displayed. Products from national mapping agencies were chosen because of their availability at different scales, their defined familial relationships and their widespread use.

The maps used were, whenever possible, obtained in digital form and, for each country, covered the same core area at the scales shown in Table 3.1. For both countries, and for all scales, the study areas were only a portion of the conventional paper map sheets. The British region under study had a rectangular shape with an area of 10 × 20 km and was one-eighth of the area covered by the corresponding Landranger map sheet at the scale of 1 : 50 000. The Portuguese region had an area of about 10 × 16 km, which was approximately one-quarter of the area covered by one IPCC map sheet at the scale of 1 : 50 000. Because of the scale reduction, for both countries, the portions of the map sheets covered by the study areas at the smaller scales were naturally progressively smaller.

The scales studied for both the British and the Portuguese study areas had a very similar range. The relatively minor differences in the case of the small-scale maps reflect the different cartographic traditions of the two national mapping agencies. For both countries, the smaller scales were considered to be generalised versions of the 1 : 50 000 scale maps, although they may have subsequently been updated separately (a detailed description of the maps' lineage is given in sections 3.1.1 and 3.1.2). This choice of scales was conditioned by the fact that the largest scale available from the Portuguese IPCC was the 1 : 50 000 (i.e. it is the IPCC basic scale map for the Sousel region).

By definition, the 1 : 50 000 scale maps of both countries were the most accurate of the maps chosen. As a consequence, the generalisation of all the smaller-scale maps was measured against the 1 : 50 000 scale map of each country. In other words, the 1 : 50 000 scale maps were taken as a baseline from which the extent of the generalisation suffered by the smaller-scale maps

Table 3.1 The range of scales used for each country.

Classification of the maps[a]	Britain; Canterbury and East Kent	Portugal; Fronteira, Sousel
Medium-scale maps	1 : 50 000	1 : 50 000
Small-scale maps	1 : 250 000	1 : 200 000
	1 : 625 000	1 : 500 000

[a] According to Maling (1989), large-scale maps have a scale larger than 1 : 12 500; medium-scale maps have a scale between 1 : 13 000 and 1 : 126 720; and small-scale maps have a scale between 1 : 130 000 and 1 : 1 million.

could be determined. It was assumed that even though the 1 : 50 000 scale maps already contained some generalisation of the ground truth, this generalisation was less than that of the smaller-scale maps. Experiments to test this assumption in the case of the British study area were carried out and can be found in section 4.3. This was done by comparing part of the British 1 : 50 000 scale map with a 1 : 10 000 scale map sheet. It was not possible to do the same for the Portuguese maps, as their mapping agency does not produce a scale larger than the 1 : 50 000 (with the exception of a few 1 : 10 000 scale maps for the city of Lisbon). It was also more important to make this comparison for the British maps, because the OS 1 : 50 000 is *derived* from larger-scale maps, while the 1 : 50 000 scale Portuguese map is the *basic* scale map. An extra map was therefore used only in the case of the British study area. The specific 1 : 10 000 map sheet was chosen so as to include portions of most of the features analysed in the other maps.

The choice of the Portuguese region was partly based on the minimum difference in years between the publication of the maps at different scales. The region of Sousel had the smallest time difference (seven years) among the available map scales. In Britain, maps are updated more frequently and, therefore, minimisation of the time difference between them was not as important as for the Portuguese maps. All of the Portuguese maps were obtained in paper form and had to be digitised. The British region was chosen because most of the map scales were already available in a digital form. The region of East Kent is part of a larger region for which the 1 : 50 000 scale map was available as a trial digital data product. This meant that only the OS 1 : 10 000 scale map and the Portuguese maps needed to be digitised (this is described in section 3.2).

Within both regions, it was important to select study areas that would share some of the same type of geographical features. This made it possible to compare generalisation effects for similar features across the different countries. The regions surrounding the town of Fronteira and the city of Canterbury were selected as in both cases there is a major river (retained at the smaller scales) and a railway that crosses the study area. While there would be differences between the two areas because of topography (e.g. the Portuguese study area had a rougher terrain and had more rivers) and different cartographic practices (e.g. in cases of conflict the IPCC displaces roads before rivers, while the reverse happens at the OS), the use of two different study areas was nevertheless useful in highlighting some of the different generalisations inherent in maps.

The lineage of the maps used is crucial in understanding the transformations that all of the maps went through during the cartographic process and to identify their different sources (e.g. aerial photographs, large-scale maps, etc.). The general classification of the maps used in this study is given in Table 3.2. Where the maps obtained were in paper form, digitising was necessary, and this was carried out by the author using raster scanning and subsequent

Table 3.2   A general classification of the spatial data obtained for this study.

Maps	Source	Format in which data were obtained	Analogue maps used to derive the digital data
UK			
1 : 10 000	OS[a]	Paper map	OS TR 15 NW
1 : 50 000	OS	NTF,[c] V1.1, level 3	OS Landranger maps 178, 179 and 189
1 : 250 000	OS	NTF, V1.1, level 3	OS Routemaster sheet 9 (South-East England)
1 : 625 000	OS	NTF, V1.1, level 3	OS maps: Routeplanner, administrative and topographic
Portugal			
1 : 50 000	IPCC[b]	Paper map	IPCC sheet 32-D (Sousel)
1 : 200 000	IPCC	Paper map	IPCC sheet 6
1 : 500 000	IPCC	Paper map	IPCC map of Portugal

[a] OS, Ordnance Survey, Great Britain.
[b] IPCC, Instituto Português de Cartografia e Cadastro, Portugal.
[c] NTF, National Transfer Format (now British Standard 7567).

semi-automatic digitising (see section 3.2). The third column of Table 3.2 shows the format in which the data were obtained and therefore specifies the cases in which digitising was necessary.

Information is also given in Table 3.2 about the specific paper maps that were used to create the digital data of both study areas. Information relating to the particular map sheets is described in detail below for the OS maps and the IPCC maps. This includes the compilation method for each map, the date of the original analogue maps, digitising procedures and any major revisions that have been carried out.

3.1.1   The British maps

The main map series produced by the OS are the following: 1 : 1250, 1 : 2500, 1 : 10 000, 1 : 25 000, 1 : 50 000, 1 : 250 000 and 1 : 625 000 scale maps. Except for scale, the 1 : 10 000 and 1 : 25 000 maps are virtually the same; the latter is re-drawn from the former to a different specification. The relationship between the different map series produced by the OS is shown in Figure 3.1. Whatever the origins of the different maps, large changes to the content have occurred over the years as they have been updated.

The three largest-scale map series (i.e. 1 : 1250, 1 : 2500 and 1 : 10 000) constitute the basic scale maps produced by the OS. All of the other scales are

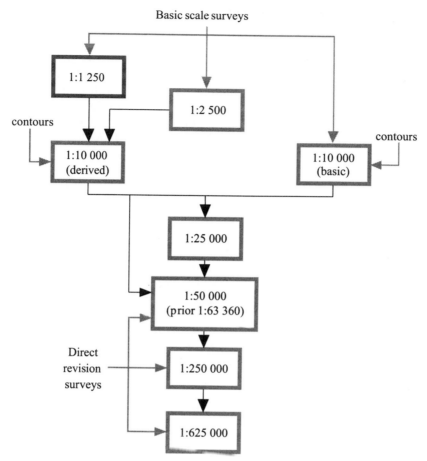

Figure 3.1 The links between the different Ordnance Survey map series (based on Department of the Environment, 1987 and Harley, 1975). The pale lines indicate survey input.

considered to be derived map series and result, at least partially, from generalisation of the former. The topographical surveys used to create and update the three basic scales are a combination of aerial photography and ground survey methods. Because the Canterbury area is mainly urban, the 1 : 10 000 sheet used in this study is a derived map, as it is partly derived from larger-scale surveys at the scales 1 : 1250 and 1 : 2500. The contours, however, result from photogrammetric surveys at the 1 : 10 000 scale and were added separately to the derived planimetric map detail. Although the 1 : 50 000, 1 : 250 000 and 1 : 625 000 scale maps were originally derived from larger-scale maps, they may have subsequently been updated separately by direct revision surveys; for example, when a new road was built. The 1 : 50 000 map in particular has a complex lineage, as it was derived initially from the older 1 : 63 360 scale map by photomechanical enlargement and recompilation (in the case of the First Series) and rescribing (in the case of the Second Series). The 1 : 50 000 Second Series obtained new information from the 1 : 10 000 scale maps (such as the metric contours) and the 1 : 63 360 map series was derived from larger scales such as the 1 : 10 000 and 1 : 25 000. Therefore, when in this study the 1 : 50 000 scale map is said to be derived from larger scales, it is meant that it derives from them both *directly* and *indirectly* (via the 1 : 63 360 map series). Notwithstanding the complex lineage of the 1 : 50 000 scale map, experiments (see section 4.3) show that this map is planimetrically accurate.

The scales 1 : 250 000 and 1 : 625 000 are dependent on the same information sources as in the production of the 1 : 50 000 scale map, although the two smaller scales contain some additional information and are produced on a more frequent revision cycle than the parent series (Smith, 1980). Because the different scales have different revision cycles, the links between the different map series are weaker than in the case of the Portuguese map series. The revision needed for both the 1 : 250 000 and the 1 : 625 000 scale maps (e.g. roads are updated annually) is applied digitally to both scales separately. The updating of the 1 : 50 000 scale map (on a cycle between one and seven years depending on how well the maps sell) is done manually by scribing.

The 1 : 250 000 scale map was derived initially from the generalisation of the old Seventh Series at the scale of 1 : 63 360. The adopted method of drafting the 1 : 250 000 scale map was to 'generalise separately the details of contours, water, and of outline at 1 : 63 360 scale, followed by redrawing and then reduction to 1 : 250 000 scale' (Harley, 1975, p. 132). The extent of generalisation and elimination of selected detail in each case was conditioned by the cartographer's aim to achieve a satisfactory representation of the landscape. Although this representation was often determined by guidelines, because of the variation of geographical features this was not always the case. Sometimes manual generalisation decisions are governed by *best depiction* rather than by a *particular rule* (Geoff Johnson, OS, 1993, personal communication).

The 1 : 250 000 and 1 : 625 000 scale maps not only required considerable simplification and generalisation of the detail of the parent map, but also deliberate emphasis of some features at the expense of others. In the case of the 1 : 250 000 scale map, the major relief patterns and the road network were enhanced, because the main aim of this map series was to provide a map that was especially useful to motorists (Harley, 1975). Although many generalisation decisions are tailored to individual cases, because the two smaller-scale maps give predominance to roads, these features should be the ones that maintain the highest level of positional accuracy. In addition, water features are generally moved first in a situation of conflict – unless they coincide with boundaries, in which case other features are displaced first (Geoff Johnson, OS, 1993, personal communication).

This higher accuracy of the roads was confirmed by an unpublished experiment carried out by the OS, in which the 1 : 250 000 scale map was enlarged and compared with the parent 1 : 50 000. The researchers found a very good fit for the major roads. The study also found that the old road network (i.e. the roads that had been generalised manually) was more accurate than the more recently updated information (Geoff Johnson, OS, 1993, personal communication). The greater discrepancy of the more modern roads came about because the updating process aims to add new information which is primarily *topologically* correct rather than *positionally* accurate (Neil Smith, OS, 1993, personal communication). When updating the smaller scales, the cartographer adapts the new information to the existing data and is therefore more concerned with maintaining the relative accuracy of the features, rather than the absolute accuracy of the features in relation to the OS National Grid.

The relevant cartographic information for all of the OS maps used in the study is summarised in Table 3.3. This includes the date and origin of the analogue source data, the digitising procedure and the digital update date. Information about the digitising procedure encompasses whether the maps were automatically or manually digitised, the date when the data were first fully digitised, and who carried out the digitising.

The 1 : 10 000 scale sheet was obtained in paper format and was digitised by the author. Only eight features were digitised from the 1 : 10 000 scale map. These were the ones needed for comparison with the same features of the 1 : 50 000 scale map (see section 4.3). The digital 1 : 50 000 trial data used in this study cover a 20 × 20 km area of Kent (tile number 6014) drawn from OS Landranger map sheets 178, 179 and 189. The data were captured by Laser-Scan (commissioned by the OS) from film positives of several map components (e.g. orange road mask, rights of way, names, etc.) which were edge-matched between individual map sheets. These trial digital data were first fully digitised in October 1990. The contour data were taken from the OS contour data sets captured by the former Directorate of Military Survey. For the digitising, Laser-Scan used VTRAK for most of the map. This same system was used by the author to digitise the OS 1 : 10 000 scale map and all of the Portuguese maps. VTRAK was crucial to this study, and a more detailed description of the system is given in section 3.2.

The 1 : 250 000 scale digital data were first manually digitised by Readers Digest Associates Limited between 1986 and 1988. The OS enhanced and released these digital data in 1990. Since then, it has updated the data yearly. The source of the digital data was the OS 1 : 250 000 scale Routemaster map series, although the OS has enhanced the digital data considerably (Ordnance Survey, 1990). The 1 : 625 000 scale digital data were manually digitised by the OS between 1977 and 1979. The first release of these digital data were in 1982, and since then they have been updated each year. The main source of the digital data was the OS 1 : 625 000 scale Routeplanner map series. From this map series, data related to hydrography, transport and urban areas were obtained.

Table 3.3 A cartographic classification of the Ordnance Survey digital maps (based on information obtained from the Ordnance Survey and from Harley, 1975).

Digital maps	Original analogue maps	Digitising procedure	Digital update date
1 : 10 000	Derived map compiled from larger-scale surveys (1 : 1250 and 1 : 2500) dated between 1955 and 1982; contours surveyed 1972; revised for selected change 1990.	Digitised by the author 12/1992, using VTRAK[a]	—[b]
1 : 50 000	Derived from the old 1 : 63 360 series which in turn was derived from the 1 : 10 000 and 1 : 25 000 scale maps. It has direct revision surveys. Derivation of contours from contours surveyed at 1 : 10 000 scale with 10 m vertical interval. Sheet 178 – revised 1979; selected revision 1983, 1985, 1987 Sheet 179 – revised 1978; selected revision 1981, 1985, 1987 Sheet 189 – revised 1978; selected revision 1982, 1985	Digitised by Laser-Scan 10/1990 using VTRAK	—[b,c]
1 : 250 000	Generalised from 1 : 50 000 scale maps. Sheet 9 – revised to 1985	Manually digitised by Readers Digest Associates Limited 1986–8; OS first release 1990, yearly updates	May 1991
1 : 625 000	Generalised from 1 : 250 000 scale maps. Routeplanner – revised to 1976 Administrative – revised to 1976 Topographic – revised to 1963	Manually digitised by OS 1977–9; first release 1982; yearly updates	May 1991

[a] Semi-automatic line-following digitising software produced by Laser-Scan Ltd.
[b] There has been no digital update of the data and therefore the digital date corresponds to its paper counterpart.
[c] This digital product is no longer supplied by the OS and has been replaced by a colour raster of the whole country at this scale.

The Routeplanner map is regarded as the base to which data from other sources have been adjusted (Ordnance Survey, 1987). These other sources are the OS 1 : 625 000 administrative map and the OS 1 : 625 000 general topography map, for additional river information. The digital data do not contain more information than is contained in these three source documents.

3.1.2 The Portuguese maps

The topographic maps produced by the civilian mapping agency of Portugal – the Instituto Português de Cartografia e Cadastro (IPCC) – constitute a single main family formed by the 1 : 50 000, 1 : 100 000, 1 : 200 000 and 1 : 500 000 scale maps. The 1 : 100 000 scale map was not used in this study, as there was no equivalent scale for Britain. The 1 : 50 000 scale map is the basic scale map produced by the IPCC. The other smaller scales are map series derived from this one.

The lineage of the Portuguese maps is shown in Figure 3.2. The 1 : 200 000 scale map results indirectly from the generalisation of the 1 : 50 000 – the 1 : 200 000 scale map is actually derived by the generalisation of the 1 : 100 000 scale map, which in turn derives from the generalisation of the 1 : 50 000 scale map (i.e. a cascading system). Besides generalisation, there is also a projection change between the 1 : 100 000 and the 1 : 200 000 scale maps. Both the 1 : 50 000 and the 1 : 100 000 scale maps use the Bonne projection and the Bessel spheroid, while the 1 : 200 000 and 1 : 500 000 scale maps use the Gauss–Krüger projection and the International spheroid (see section 3.3). The 1 : 500 000 scale map resulted from the generalisation of the 1 : 200 000 scale map. The 1 : 200 000 and 1 : 500 000 scale maps were exclusively derived by generalisation of the larger-scale maps and no fieldwork was carried out. Throughout the whole process, the placing of the triangulation points was the most precise operation. Across all of the scales, the positions of the triangulation points were fixed and therefore were not affected by generalisation.

The 1 : 50 000 scale map was obtained directly from 1 : 30 000 scale aerial photographs using a stereo-plotting machine, with the operator drawing the

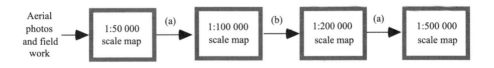

(a) generalisation

(b) generalisation and projection change

Figure 3.2 The origin of the families of maps of the IPCC.

different details (e.g. roads, rivers and buildings) in relation to the control points. These control points included the triangulation points and other photogrammetric points which could be easily identified on the ground (e.g. a road crossing). IPCC topographical surveys, needed to determine the exact location on the ground of these control points, are usually a combination of aerial and ground survey methods. The updating of the 1 : 50 000 scale map series was done by overlaying recent 1 : 15 000 scale aerial photographs on top of the old versions of the paper maps. By zooming into the areas that needed updating, to a scale of 1 : 12 500, the cartographer could draw with precision the detail of new or modified features. If maps are too out of date (some of the IPCC maps date back to the 1950s), they are re-drawn from aerial photographs using stereo-plotting techniques (Guedes, 1988). The production of the 1 : 50 000 scale map can also make use of other sources depending on their availability for the different regions and how current those sources are. These sources can include orthophotomaps (mainly at the scale of 1 : 10 000), and the 1 : 2000 and 1 : 5000 large-scale maps in regions where they are already available. When generalising from these two very large scales to the 1 : 50 000 scale map, the generalisation was carried out to an intermediate scale and then the map was reduced photographically (Instituto Português de Cartografia e Cadastro, 1988).

For the generalisation of topographic maps and their elements such as roads, settlements and hydrographic features, the IPCC uses as a general reference the rules of the Swiss Society of Cartography (Swiss Society of Cartography, 1987) (José Guedes, IPCC, 1992, personal communication). In their publication, the Swiss Society of Cartography explains what they consider to be acceptable ways of generalising topographic maps, by illustrating good and bad examples of generalisation. There are, in addition, some IPCC in-house rules relating to the order in which features are displaced in cases in which conflict for space occurs. The order of displacement (starting by the features that are displaced first) is as follows: green areas, houses, roads, rivers/streams and railways (José Guedes, IPCC, 1992, personal communication). This means

Table 3.4 A cartographic classification of the Instituto Português de Cartografia e Cadastro maps.

Digital maps	Original analogue map	Digitising procedure
1 : 50 000	1975 (second edition)	Digitised by the author in 1991, using VTRAK[a]
1 : 200 000	1974 (first edition), based on 1970 road information	Digitised by the author in 1991, using VTRAK
1 : 500 000	1974 with 1981 revision (second edition)	Digitised by the author in 1991, using VTRAK

[a] Semi-automatic line-following digitising software produced by Laser-Scan Ltd.

that green areas are the first to be shifted and the railways are the last to alter position. In general, therefore, railways should be the features with the most accurate position, except for the location of the triangulation points (which have a fixed position). This is the same order as observed by Dutch cartographers (Müller and Mouwes, 1990). The relevant cartographic information for all of the IPCC maps used in the study, such as the date and origin of the analogue source data and the digitising procedure, is summarised in Table 3.4. All of the Portuguese maps were obtained in a paper format and were digitised by the author. This operation is described in the next section.

3.2   Converting paper maps into a digital form

The digitising by the author was done using Laser-Scan's powerful VTRAK software – an interactive line-following digitising method (Laser-Scan, 1992). VTRAK uses scanned map data and a range of algorithms that allow for a precise capture of the detail of the different features. Features are automatically collected in vector format, but the process can be stopped at any time when an error occurs or when automatic line-following is interrupted due to graphic ambiguities or bad line quality. When parts of lines cannot be followed automatically (e.g. due to lettering, other lines, gaps on the line, etc.), VTRAK allows for the operator interactively to introduce manual points or jump over features. The software also allows the captured data to be edited and verified in direct comparison with the original scanned map.

An important concern in converting the paper maps into a digital form was that the process chosen was as accurate as possible and that, at the same time, would not contribute to further generalisation of the maps. This requirement was instrumental in the choice of VTRAK for data collection. Because it uses semi-automatic line-following techniques, it is capable of higher-accuracy data capture than manual digitising methods (Thapa and Boosler, 1992) and is less dependent on the individual operator. Devereux and Mayo (1990), for example, compared conventional manual digitising with semi-automatic line-following techniques and found that the latter generated results which were 37 per cent less variable than the former. This means that the same line digitised at different times or by different operators using semi-automatic line-following techniques, such as VTRAK, will be very similar. Compared with manual digitising, semi-automatic line-following techniques are more accurate, as well as being more reproducible and consistent methods. As a consequence, these techniques can capture features in a more comparable fashion, and the resulting digital maps are the closest possible representation of the original analogue maps.

The performance of VTRAK depended on the values of the parameters used. The parameter files of VTRAK need to be tuned to the specific map being digitised. When tuning, care was taken in controlling the parameters that influenced line characteristics. The first step to accomplish this was to classify the parameters used by VTRAK and to identify the ones that affected line characteristics. For the purpose of this study, the parameters used by VTRAK were classified into two main types: parameters of line-following and of point distribution.

The line-following parameters should be altered to conform with the characteristics of the different lines, such as line width, line quality and angular corners. The point distribution parameters are those that will influence line characteristics the most, and therefore they were kept similar for the different scales. The pitch level, which determines the number of times VTRAK scans a particular line, should vary according to the scan resolution but, as all maps were scanned with the same resolution (50 μm), all maps were digitised at the same pitch level. Besides the parameters that affected line-following and point distribution, there are other parameters that had to be kept carefully tuned, such as the speed at which VTRAK follows the lines. For this study the speed was reduced to the minimum in order to make it easier to check the quality with which VTRAK followed the lines.

Accuracy checks were carried out during and after the use of VTRAK. The vector files resulting from the use of VTRAK were always edited with the scanned maps in the background. When checking line characteristics, the digitised map was displayed on top of the raster image (i.e. the scanned map) and visual checks were made on all lines in each map. In this way, the editing was controlled by the original image, which made it easier to detect and correct errors. Small parts of the maps which did not lend themselves to efficient line-following using VTRAK (due to bad quality of the scanned map or excessive concentration of features) were digitised manually on the screen. Plots were produced and checked to ensure quality control and editing.

The scale at which the different features were digitised (i.e. the scale after enlarging on the screen) was also an important factor in assuring the digitising quality. In most circumstances, the larger the scale used for digitising the maps is, the more accurate the digitising can be. Not only can the performance of VTRAK be checked more carefully, but also, when it was necessary to add points manually on the screen this could be done with greater accuracy (the equivalent to digitising manually with a powerful magnifying glass). The editing of the maps with a raster backdrop was also done using very large scales (from approximately 1 : 4000 to 1 : 9000). The larger scale was for the digitising of intersection points (e.g. between a road and a river) with high positional accuracy. The digitising of the triangulation points was done on the screen using an even larger scale of approximately 1 : 1000, as these points had to be the most accurately digitised.

When the parameter file is well tuned for the specific map being digitised, then the accuracy of VTRAK depends mainly on the resolution of the scanned map. VTRAK detects the position of lines on the original document by comparing changes between pixel values. Because of the definition of a pixel, this position can only be accurate to half a pixel (Laser-Scan, 1992). As all maps were scanned with a resolution value equal to 50 μm, the *best* possible digitising accuracy was equal to 0.025 mm. Conversely, for the worst case scenario, the *lowest* accuracy value for each individual line was equal to half the line width, because at worst a point would be placed on the edge of the line rather than on the centre. When using VTRAK, or when digitising manually on the screen, all of the points were always within the band of the line pixels. This was ensured by the powerful magnification of the scanned image, the correct choice of parameters, and care taken in the digitising.

When features were represented by a double line, the centreline was digitised instead. The important factor in this case was the distance between the lines rather than the line width. The line thicknesses for the different features for all scales and the clearance between parallel lines (when appropriate) in millimetres of map sheet, as well as an estimate of the worst case digitising accuracy for each feature in ground units, are shown in Table 3.5. The digitising accuracy was associated with the line thickness or, in the case of features represented by a double line, the clearance between the lines. Thicker features had a potentially lower value of positional accuracy when digitised. Features represented by double lines, especially when widely spaced, were the features that could be the most imprecisely digitised (see Table 3.5). In the case of the 1 : 10 000 scale map, many of the features were represented by double lines and *to scale*, which meant that the distance between the parallel features could be even larger than the values presented in Table 3.5, and therefore the corresponding worst case accuracy values could also be larger.

In most cases, the digitising accuracy of VTRAK was better than the worst case figures shown in Table 3.5. A line digitised by VTRAK was most likely to coincide with the centre of the true line. In other words, there would be more points closer to the actual location of the line than further away from it. When a line is digitised many times using a digitising method such as VTRAK (as in the study by Devereux and Mayo, 1990), the distribution of points along the line will approximate the normal distribution, with the peak centred on the midpoint of the line (a concept closely related to the 'epsilon band' used by Blakemore, 1983). As a consequence, line-following techniques should cause fewer digitising errors than manual table digitising. An unpublished trial conducted in 1989 by Laser-Scan, using OS 1 : 1250 scale data, found that the average digitising accuracy of VTRAK was equal to 0.052 mm. This is much smaller than the manual digitising error of 0.25 mm quoted by Petrie (1990), although Bolstad *et al.* (1990) found in their study that the average uncertainty associated with manual digitising was as low as 0.054 mm. From the above, and considering the care taken in digitising, the estimated accuracy of the VTRAK digitising carried out for this study was better than 0.06 mm.

The Portuguese maps were *initially* scanned in grey scale. This meant that the pixels of the scanned images had a grey tone, which depended on the original colours of the paper maps. The use of a spectrum of grey tones,

Table 3.5   Line thickness and digitising accuracy for the different features by scale.

	Gauge of individual lines (mm)	Clearance between parallel lines (mm)	Worst case digitising accuracy (m)[a]
IPCC 1 : 50 000 (scanned at 50 μm)			
Main rivers	0.25	0.5[d,e]	12.5
Minor rivers	0.1–0.25[b]		2.5–6.2
Major roads	0.23–0.4[c]	0.3	7.5
Municipal roads	0.18	0.4	10.0
Minor roads	0.14	0.5	12.5
Boundaries	0.17		4.2
Railways	0.54		13.5
IPCC 1 : 200 000 (scanned at 50 μm)			
Rivers	0.1–0.25[b]		10–25
Major roads	0.50		50
Minor roads (double line)	0.15	0.4	40
Minor roads (single line)	0.30		30
Boundaries	0.20		20
Railways	0.54		54
IPCC 1 : 500 000 (scanned at 50 μm)			
Rivers	0.08–0.14[b]		20–35
Major roads	0.66		165
Secondary roads	0.37		90
Municipal roads	0.21		50
Boundaries	0.17		40
Railways	0.28		70
OS 1 : 10 000 (scanned at 50 μm)			
Rivers	0.18	0.64[e]	3.2
Classified roads	0.18	1.06[e]	5.3
Named roads	0.18	1.27[e]	6.4
Boundaries	0.50		2.5
Railways	0.60		3.0

[a] Equal to half the line width, or half the distance between parallel lines in the case of features represented by a double line.
[b] The Portuguese rivers have a variable gauge, as they are thinner at the source and get thicker as they approach the mouth.
[c] The major roads are represented by a double line of different thickness.
[d] Only the main river is drawn as a double line.
[e] Drawn to scale if the features are wider.

Figure 3.3   The procedure used by the author for converting the Portuguese paper maps to digital form. The numbers marked on the boxes refer to the different phases discussed in the text.

Colour paper maps:
1:50 000; 1:200 000 and 1:500 000

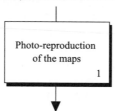

Photo-reproduction
of the maps

1

B/W acetate maps with rivers enhanced
and intensity of green areas reduced

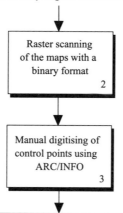

Raster scanning
of the maps with a
binary format

2

Manual digitising of
control points using
ARC/INFO

3

Semi-automatic digitising of the maps:
• structure the raster images
• enter control points in mm of map sheet
• set optimising parameters
• interactive line following with VTRAK
• post-processing
• editing with Mapstation and LITES2   4

Map transformations

5

Correction of
systematic error

6

instead of only black and white, can greatly improve the efficiency with which VTRAK follows the lines by enhancing particular features or suppressing background noise. Although most of the line features could be isolated by their specific grey tone, there was a problem with the forest areas. This was because the green colour of the forest areas had a very similar grey tone to the blue representing the rivers and, therefore, when structuring the images to attenuate the forest areas, the rivers would also disappear. There was no threshold available that would remove the forests while retaining the rivers. The solution was to photo-reproduce the original paper maps using coloured filters to attenuate the green of the forest areas and enhance the blue of the rivers. The complete process that was used to convert the Portuguese paper maps into digital form is shown in Figure 3.3.

As a result of the photo-reproduction (phase 1 in Figure 3.3), the maps created were black and white and, compared with the original colour maps, the rivers were enhanced. Because the maps were now in black and white, they were scanned with a compressed binary format (phase 2) and not in grey scale (this had the advantage of taking up ten times less file space than when the maps were scanned in grey scale). Also, by carrying out the photo-reproduction process the paper maps were converted into acetate, which avoided any further potential paper deformation. When digitising the OS 1 : 10 000 scale map, phase 1 was not necessary. The 1 : 10 000 scale map was black and white (except for the contours which were brown, but did not need to be digitised) and therefore the scanning was carried out in binary rather than grey scale.

Although care was taken to obtain new and unstretched paper maps, some paper deformation probably occurred. To correct for possible paper deformation, the control points of the Portuguese maps and of the OS 1 : 10 000 scale map were entered in millimetres of map sheet. These values were then transformed to the appropriate coordinates at a later stage (phase 5 in Figure 3.3). This transformation from millimetres to the coordinates of the different control points corrected for any existing paper deformation. Rather than measuring with a ruler, in order to determine the values in millimetres, the control points of the paper maps were manually digitised using a digitising table and the distance between them calculated (phase 3). Once the positions of the control points were determined in millimetres of map sheet, the maps were ready to be digitised.

As mentioned earlier, the semi-automatic digitising of the maps was done using VTRAK (phase 4 in Figure 3.3). First, the raw scanned data had to be converted into Laser-Scan's Structure Image format. This reorganisation of the raster scanned image data is essential before VTRAK can extract any features. As with any other form of digitising, the next step was to enter the control points, in millimetres of map sheet. Before starting the interactive feature extraction with VTRAK it was also necessary to edit the file that contains the set of parameters that control the line-following capabilities. After all the data were digitised, some post-processing followed, without which the data could not be manipulated. Finally, the digital data were edited using LITES2 and Mapstation with the scanned map in the background. Two more stages were needed to convert the Portuguese paper maps into digital form. These were map transformations and the correction of systematic error (phases 5 and 6 respectively), which are described in more detail in the next two sections.

3.3   Map transformations

In the case of the maps digitised by the author (i.e. all of the Portuguese maps and the OS 1 : 10 000 scale map) it was first necessary to transform the control points of the maps from millimetres of map sheet into the appropriate coordinates. As explained in the previous section, this transformation was carried out to correct any potential paper deformation of the maps. Unlike the Portuguese maps, the British maps did not need any other transformation, because all of the British maps had the same projection, spheroid and datum.

For historical reasons, the Portuguese 1 : 50 000 scale map is made using the Bonne projection and the Bessel spheroid, while the more recent 1 : 200 000 and 1 : 500 000 scale maps use the Gauss–Krüger projection and the International (or Hayford) spheroid. In order for the Portuguese maps to be comparable, they had to be transformed to the same projection and spheroid. The different transformations performed on the three Portuguese maps are shown in Table 3.6.

In both the British and Portuguese cases, the projection used in the transformation was the Universal Transverse Mercator (UTM), so as to take advantage of metric units. Although, by definition, the use of geographical

Table 3.6   Map transformations used for the Portuguese maps.

	Scales		
	1 : 50 000	1 : 200 000	1 : 500 000
Original			
Projection	Bonne	Gauss–Krüger	Gauss–Krüger
Spheroid	Bessel	International	International
Datum	Lisbon	Lisbon	Lisbon
Final			
Projection	Transverse Mercator	Transverse Mercator	Transverse Mercator
Spheroid	International	International	International
Datum	Lisbon	Lisbon	Lisbon

coordinates would eliminate most distortion, the UTM was preferable because it uses metric units rather than degrees, minutes and seconds. However, before proceeding with the generalisation measurements, the distortion caused by the UTM projection had to be taken into account. Any map projection that transforms features on the curved Earth's surface on to a planar one causes stretching of the mapped surface. This stretching causes variations in scale from one part of the map to another. This local scale change means, for example, that the measurement of a line represented on a small-scale map could be dependent on its position on the map, its orientation and its length. Also, the shape of features can change – a line that is straight on the ground might no longer be straight on a map. The first factor to influence the amount and variation of the local scale factors is the projection used.

The UTM projection is especially suitable for a country which has its greatest extent in a north–south direction (such as in the case of both Britain and Portugal), as it will reduce overall distortion (Harley, 1975). This is because most of the distortion occurs in the east–west direction and is far less influenced by the north–south positioning. The location of the study area in relation to the centre of the projection (i.e. the meridian of zero distortion) will determine the amount of east–west distortion: the further away from the centre of the projection, the more east–west distortion there will be. The positions of both study areas in relation to the central meridian are shown in Table 3.7. The British study area was located further away from the central meridian and therefore had a larger local scale variation.

The other factor that influenced the amount of the local scale change between the two extremes of the study area (i.e. the difference between the local scale at the eastern and western borders) was the size of the study area. Significant scale changes within a single map sheet only occur when a sufficiently large portion of the Earth's surface is shown on one map (Maling, 1989); in other words, when using very small scale or atlas-scale maps. If the portion of the Earth's surface under study is relatively small, the scale factor can be considered to be homogeneous across the study area. According to Harley (1975), local scale factors may be assumed to be constant for a small zone of about 10 km radius. Both the Portuguese and the British regions had a maximum length only slightly wider than 10 km, and so changes in the local scale factors were minimal. The British region under study had an area of 10 × 20 km (which is one-eighth of the area covered by one Landranger map sheet at the scale of 1 : 50 000) and the Portuguese region had an area of 10 × 16 km (which is approximately one-quarter of the area covered by one IPCC map sheet at the scale of 1 : 50 000).

The British region was slightly wider than the Portuguese study area (20 km rather than 16 km) and so had a larger variation of the local scale factor between the eastern and the western borders of the study area. The variation was also larger because the British region was further away from the central meridian. The eastern border of the British study area lay 3°08′20″ away

Table 3.7 The positions of both study areas in relation to the central meridian.

	Fronteira, region of Sousel, Portugal (16 km long and 10 km wide)	Region of Canterbury and East Kent, Britain (20 km long and 10 km wide)
Origin of the projection (central meridian)	+39°40′ north, 8°07′54″ west of Greenwich	+49° north, 2° west of Greenwich
Study area's western border	22′54″ east of central meridian	2°52′35″ east of central meridian
Study area's eastern border	33′54″ east of central meridian	3°08′20″ east of central meridian

from the central meridian, while the eastern border of the Portuguese study area only lay 33'54" away. The western border of the British study area had a local scale factor equal to 1.00009, while the eastern border had a local scale factor of 1.0002 (based on the local scale factors given by Harley, 1975). As a consequence, a line with a ground length of 5000 m could be 5000.5 m long on the map if it was located on the western edge of the British study area. The same length line positioned close to the eastern border could have a map length of 5001.0 m. This corresponded to a maximum 0.02 per cent change in length. Therefore, any difference caused by local scale factors (even in the worst case example of the British study area) was minimal.

Although the measurement of this distortion could be relevant for certain studies (e.g. when comparing different regions within a country), this study could be considered to be independent of this type of error. All of the maps of each study area covered the same core area, and so the distortion was the same for all of the maps of each region. Therefore, not only was the error caused by the change in local scale very small, but also the error was constant across the different map scales of the two study areas. This was crucial as, irrespective of the size of the error, the maps remained comparable. For these reasons, local scale factors *were not applied* to the measurements performed in this study.

3.4   The correction of systematic error

Following the map transformations, the systematic error of all the maps was investigated. Systematic error, as the name indicates, is considered to be a constant error that affects the map as a whole. This type of error is generated during the map creation process; for example, due to the malpositioning of the grid (Harley, 1975). For the British study area, the three scales were already obtained in a digital format and systematic error had been corrected by the OS. However, in the case of the three Portuguese maps digitised by the author, systematic error had to be corrected. The grid of the Portuguese maps, which was calculated from the triangulation points, was the last to be drawn on the maps and might have suffered some displacement. This systematic error had to be eliminated; otherwise, it would have affected the results of some of the generalisation descriptors, especially in the case of the measurement of the displacement of the different features.

In order to determine and correct the systematic error it was necessary to select points for which the true ground location was known. For this exercise, triangulation points were used, as their position is the most accurate procedure of the whole map-generating process (José Guedes, IPCC, 1992, personal communication). Triangulation points were placed in the maps using very precise automated techniques based on the geographical coordinates of the points. When the maps were digitised, the coordinates of the four corners of the maps were entered (as is common practice), but if the grid was misplaced then this would have caused a systematic error which could now be corrected by using the coordinates of the triangulation points. The distance between the location of the triangulation points on the different maps and the coordinates of their true location (obtained from the IPCC) gave the *size* of the positional error of those points. Positional error with a preferred *direction* was the tell-tale sign of systematic error (Ian Dowman, 1992, personal communication).

To correct the positional error of the base control shown on the maps, an empirical transformation was applied. The map coordinates of the triangulation points were shifted until a best fit was found with the true coordinates. In order to determine this best fit, a least squares method was required. However, a least squares transformation is designed for 'local' fitting and works better if large constant values in the coordinates are removed beforehand (Laser-Scan, 1992). Therefore, before applying the transformation, the triangulation points of the different maps were moved (using the editing software LITES2) until a *visual* best fit was obtained for their true location. Because LITES2 does not allow for the selection of multiple features, the triangulation points of each scale had to be linked to each other. An artificial line was created that linked all of the triangulation points. It was this 'aggregation' of triangulation points that was translated until the triangulation points of the different maps visually appeared to be closest to their true locations. This operation permitted the determination of the x and y translation coordinates which were used for the first transformation of the maps. This transformation eliminated a large proportion of the systematic error and improved the accuracy of the next transformation.

The next step was to use the more accurate empirical transformation to obtain the best fit between the two sets of coordinates (i.e. the true coordinates and the ones in the maps) and therefore correct the remaining systematic error. From the different transformations available, an orthogonal transformation (rather than a curvilinear transformation that would warp the map) was selected. This transformation only corrects for rotation, scaling and translation. The important implication of this is that the shapes of features are not altered and that all of the features in the map (independent of their relative position in relation to the triangulation points) are affected in the same way. This transformation uses a minimum of two points and a maximum of 20, and its effectiveness is independent of the number and the distribution of the control points throughout the study area. This last point proved essential in the case of the smallest-scale map, which had fewer triangulation points. After the best fit was found for the triangulation points, the rest of the features were shifted using the same transformation.

The angles and lengths of the displacement vectors between the location of the triangulation points on the maps and the 'true' location as determined by topographical survey are shown in Figures 3.4–3.9. If a map had a large sys-tematic error, then all vectors had similar angles and the length of the vectors was large (i.e. the map was shifted in one direction). If, on the contrary, there was not much systematic error, then the angles of the displacement vectors had contrasting values and their lengths were relatively small. In the latter case, the differences between the locations of the triangulation points on the map and in surveyed terms were mainly due to random errors, the source of which could not be pin-pointed. The only map which did not appear to have much systematic error (see Figure 3.4) was the 1 : 50 000 scale map. There were 12 triangulation points on the 1 : 50 000 scale map within the Portuguese study area. As can be seen from Figure 3.4, the angles of the displacement vectors ranged from about 50° to nearly 350°. This spread of values suggests that the existing error was mainly random. Even so, an attempt was made to minimise the error.

The positional error of each triangulation point was individually calculated. In the case of the 1 : 50 000 scale map, before any correction the maximum displacement observed in one of the triangulation points was 30 m. However, in order to give an indication of the *mean* positional error of the maps, the root mean square error was also calculated. The root mean square error was chosen because it is an accepted and common way of evaluating an average amount of error from individual error values, and measures simultaneously both systematic and random errors (Harley, 1975; Maling, 1989). The root

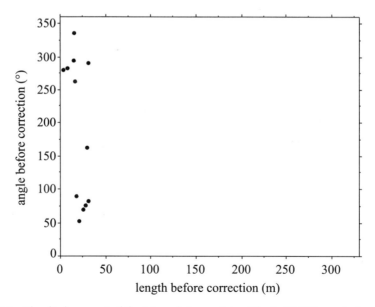

Figure 3.4   The displacement of the triangulation points of the 1 : 50 000 map before correction.

mean square error is calculated by the following formula:

$$r = \sqrt{\left(\sum_{i=1}^{n} x_i^2\right)\Big/ n} \tag{3.1}$$

where x_1, x_2, ..., x_n are the errors at n check points (Harley, 1975), which in this case corresponded to all of the triangulation points present in the maps.

The root mean square error of the length of the displacement vectors of the 1 : 50 000 scale map (before any correction was applied) was 0.42 mm, which gave an approximate accuracy of 21 m. The resulting positional error after applying the orthogonal transformation is shown in Figure 3.5. After correction, the root mean square error of the displacement vectors was reduced to 17 m (or 0.34 mm on the map). The application of the transformation meant that, although the overall error of all the triangulation points was reduced (albeit by a small amount), the displacement vectors of some of the individual points actually increased in length. For example, the maximum displacement of the triangulation points after correction increased to 31 m (cf. 30 m before the transformation).

There were only eight triangulation points, within the study area, at the scale of 1 : 200 000. The displacement vectors of the 1 : 200 000 scale map showed a different pattern from that of the 1 : 50 000 scale map, and this can be seen in Figure 3.6. The angles of the displacement vectors of the 1 : 200 000 scale map showed a preferred direction that indicated the existence of a sizeable systematic error in the map. The angles of the displacement vectors were

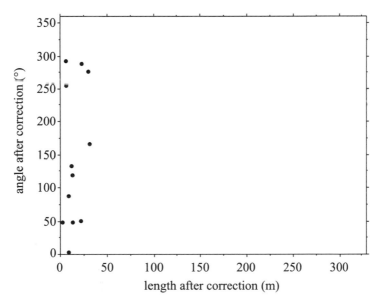

Figure 3.5 The displacement of the triangulation points of the 1 : 50 000 map after applying an orthogonal transformation.

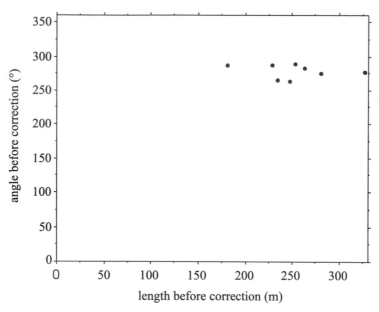

Figure 3.6 The displacement of the triangulation points of the 1 : 200 000 map before correction.

very concentrated in a narrow range of values, between 264° and 290°. The lengths of the displacement vectors were also relatively large. Most error was greater than 200 m and less than 400 m (87 per cent of the cases), which corresponded to an error on the map of between 1 mm and 2 mm. The root mean square error of the displacement was equal to 253 m, which was equivalent to 1.26 mm. The resulting angles and lengths for the same displacement vectors, after correcting the systematic error of the 1 : 200 000 scale map, are shown in Figure 3.7.

As can be seen in Figure 3.7, the lengths of the displacement vectors of the 1 : 200 000 scale map were reduced. The maximum displacement corresponded to 78 m, which was equivalent to 0.39 mm. In terms of the root mean square error, the displacement was now 54 m, which corresponded to 0.27 mm accuracy. The angles, on the other hand, now ranged from 5° to 334°. Much of the systematic error had been eliminated and the remaining error was mainly random. The last map to be corrected for its systematic error was the 1 : 500 000 scale map. The initial positional error is shown in Figure 3.8, and the remaining positional error after the transformation was applied can be seen in Figure 3.9.

It can be seen in Figure 3.8 that the angles of the three displacement vectors had very similar values and that the lengths of the displacement vectors were relatively large. The root mean square error of the displacement before correction was equal to 535 m (1.17 mm). Although only three triangulation points remained on the study area of the 1 : 500 000 scale map, it was still possible to

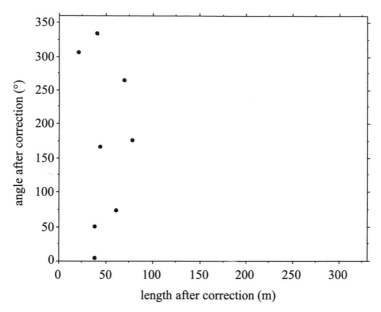

Figure 3.7 The displacement of the triangulation points of the 1 : 200 000 map after correction of the systematic error.

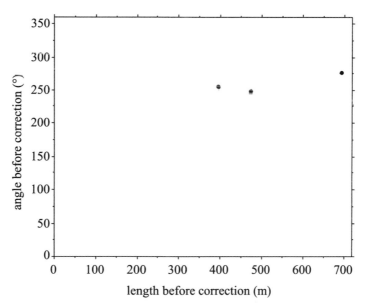

Figure 3.8 The displacement of the triangulation points of the 1 : 500 000 map before correction.

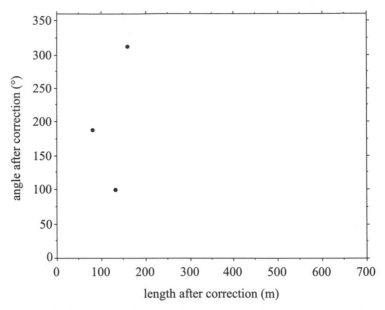

Figure 3.9   The displacement of the triangulation points of the 1 : 500 000 map after correction of the systematic error.

correct for systematic error, because only two points were needed to carry out the orthogonal transformation. Also, the three triangulation points were distributed across the map in a pattern that allowed for correction in both the north–south and east–west directions (i.e. the three points were not all in a straight line). After correction of the systematic error, the angles of the three displacement vectors differed widely, as can be seen in Figure 3.9. The maximum individual displacement of the triangulation points of the 1 : 500 000 scale map was 312 m, which corresponded to 0.3 mm, and was the smallest of all three maps. The displacement vectors now had a root mean square error of 127 m, which was equivalent to an accuracy of 0.25 mm.

The objective of these transformations was successfully to correct for systematic error. The final root mean square errors of the three maps were: 0.34 mm for the 1 : 50 000 scale map; 0.27 mm for the 1 : 200 000 scale map; and 0.25 mm for the 1 : 500 000 scale map. This error that remained was mainly due to random errors. After the correction of systematic error, the conversion of the paper maps to a digital form was finalised. However, some editing of all the maps was still necessary. This is described in the next section.

3.5   Final editing of the digital maps

The final editing of the digital maps included the clipping of the study area, the coding and linking of features, and the elimination of spurious spikes in the lines. The OS 1 : 50 000 scale trial digital data required extensive editing as a result of the vector data specifications. The rivers, streams and canals, for example, were not contiguous, and there were large gaps in water features, such as where the River Stour passed through the city of Canterbury. These gaps existed on the map base and were not closed off when the digital data were created (Ordnance Survey, 1991). For this study, it was essential that 'whole' features – rather than arbitrarily selected sections of them – were used in order that the same feature could be compared across scales. Editing was therefore necessary in order to close these gaps. The different sections of the rivers were linked with straight lines. This may have introduced small errors (e.g. on length calculations) where the missing lengths of river were not actually straight.

Other features required editing as well, although to a lesser extent than the water features. For example, the roads and railways of the British 1 : 50 000 scale map showed many interruptions. This meant that instead of just two railways and one motorway there were several fragmented components. These interruptions were caused by junctions with other roads and railways. Certain point features, such as level crossings and railway stations, also caused interruptions in the road or railway network. As a result, editing was necessary to link together all of the relevant segments. Because of the complexity of the road network of the original map data, and in order to avoid linking features incorrectly, most of this editing had to be done interactively.

It was also necessary to check for spurious 'spikes' in the lines. When digitising or even editing the lines (e.g. when linking the fragmented features of the OS 1 : 50 000 scale map), imperceptible spikes were sometimes introduced into the lines. VTRAK also occasionally introduced spikes when an intersection with another line was crossed. These spikes were usually so tiny that they had a minimal effect on most measurements, such as line length. However, for the measurement of angularity they could have caused a very large impact because, although very small, the spikes usually involved very large angular changes (in some cases up to 150°). In order to find the spikes, all angular changes larger than 45° were checked using a specially written macro. All the spurious spikes were then eliminated.

With this final editing of the digital map data, the two study areas were ready for the measurement of generalisation effects. However, it was still necessary to determine the positional accuracy of the maps, so that the total error of the maps (excluding generalisation error) could be used as a baseline when carrying out the measurements of generalisation effects. The next chapter describes how this baseline error value was calculated for the maps of both countries, how the generalisation measurements were made, and also how practical problems in carrying out the measurements were overcome.

CHAPTER FOUR

A methodology for the measurement of generalisation effects

In this chapter, the method used to evaluate the effects of generalisation within a GIS is described. This included measuring quantitatively, across scales, the generalisation effects embedded in map data for different features, comparing manual and automated generalisation procedures, and carrying out two commonly used GIS map manipulations relying on different source-scale data. Of key importance was to clarify when map error was due to generalisation effects or when it was due to other sources of error. For this, it was necessary to calculate the final positional accuracy of the digital maps of both study areas. On the basis of this final error, a novel method is proposed of distinguishing generalisation effects from other errors by the use of a well defined 'cut-off point' above which error is more likely to be due to generalisation. Past studies have always had difficulty in untangling non-generalisation map error from generalisation error.

A vital part of the methodology was the application of a GIS to carry out the measurements themselves. Using GIS as a research tool allowed a large number of different measurements to be carried out: some of the measurements would be too time-consuming or not even practical (such as the sideways shift of the lines) to be done manually. The use of a GIS also meant that the study was more reproducible. Many of the differences portrayed on maps of different scales can be visually observed and, therefore, can be easily evaluated *qualitatively*. However, the same is often not true for *quantitative* measurements of map differences. Even the manual counting of all features appearing on subsections of a map is a difficult and time-consuming task. GIS facilitates these tasks by measuring features rapidly and accurately. The way in which GIS performed these measurements ensured they were all done using the same procedure, and that they remained comparable.

For any one map, a large number of measures, or *descriptors*, of generalisation can be applied. A number of these descriptors were selected; these were taken partially from previous research on line generalisation, but a few were also especially developed for this study. Because such a comprehensive study on the quantification of generalisation effects had not been carried out before (see section 2.2), this study was to a certain extent exploratory. As a result, some of the measurements made of generalisation descriptors were subsequently rejected because they were found to be unsuitable, or were adapted to give more satisfactory results. The methodologies developed and tested for this research should be useful for further studies on generalisation.

The process starts by calculating the positional accuracy of the maps in order to distinguish between generalisation and non-generalisation error. A description of the development, selection and implementation of the different generalisation descriptors then follows. The generalisation of the British 1 : 50 000 scale map is evaluated in order to validate measuring the generalisation effects in relation to the 1 : 50 000 scale map rather than using a larger scale. An explanation of the procedure used for determining the fractal surfaces of both study areas in terms of the roughness of terrain then follows. The chapter concludes with a description of the GIS map manipulations carried out, and the methodology used for measuring generalisation effects that result from the Douglas–Peucker generalisation algorithm.

4.1   Separating generalisation effects from other sources of error

In order to measure generalisation effects, it was crucial to try to separate them out from other sources of error. In this section, a cut-off point is calculated above which the error is *more likely* to be due to generalisation rather than other sources of error. This was particularly important for the analysis of feature displacement. In map-making, much effort is expended in ensuring that a few key points (such as triangulation points) are placed as accurately as possible across all map scales. However, many features are shifted from their true positions not only due to generalisation but also due to 'non-generalisation errors' associated with the production of the source map and from digitising.

It is assumed that large *gross errors* have been eliminated from the data, as there is a very low probability that a gross error would remain undetected throughout the map-making process when periodic corrections are made (Neil Smith, OS, 1993, personal communication). *Systematic error* (a constant error that affects the map as a whole) was deliberately eliminated from the maps of both study areas (see section 3.4). Following the elimination of the systematic error, and assuming that there were no gross errors, the remaining non-generalisation error can be considered to be due solely to *random errors*.

In order to analyse the displacement of the features, it was therefore necessary to clarify when displacement of the features was caused by random errors
and when it was due to generalisation. Random errors are usually small
(Thapa and Bossler, 1992) and large gross errors are very rare. Therefore, as
displacement of features increases (above a defined cut-off point) it is more
probable that the displacement is due to generalisation. In order to determine
this cut-off point between generalisation and non-generalisation error, it was
necessary to estimate the positional accuracy of the maps excluding generalisation.

The total positional accuracy of the digital maps is a combination of errors
arising from the positional accuracy of the source material (such as drawing
and registration accuracy) and errors due to the digitising process. It is difficult to assess this total error, because the functional relationship amongst the
various errors is unknown (Thapa and Bossler, 1992); in other words, there is
a possibility that the errors in individual features at each stage of the map-
making process can cancel each other out, in direction and magnitude. When
different sources of error combine independently, and assuming a linear
relationship between the total error and the individual errors, the total error
can be *estimated* by combining the individual errors of the different map production stages using the following formula (NCGIA, 1989; Thapa and Bossler,
1992):

$$E_{\text{total}} = \pm \sqrt{\sum_{i=1}^{n} e_i^2} \tag{4.1}$$

where E_{total} is the total error of the map, and $e_1, e_2, ..., e_n$ are the individual
errors (e.g. in RMSE terms – see below) of the n different stages in the production of the map data.

The positional accuracy of a map might change for different features and
for different points on the map. It is very unlikely that the positional accuracy
of a map will be homogeneous across the whole map sheet or map tile. Many
maps created traditionally are now updated using modern procedures and, as
some features change more frequently than others, this updating process might
cover only specific sections of the map. Because of this, a particular map sheet
can be the result of a combination of different map-making procedures with
varying accuracies. The calculated positional error of the maps can therefore
only be an estimation, and it can only be an average value for the whole map.

In this section, the individual error values of the different map production
stages are expressed in terms of the Root Mean Square Error (RMSE; see
section 3.4 for the RMSE formula). The main shortcomings of the RMSE are
that it can only be calculated for readily identifiable points, and the RMSE
figure itself gives no indication of the frequency distribution of the errors. 'The
measure of absolute positional accuracy should be both a mean error and a
standard deviation, so that the maximum likely error can be estimated' (Smith

and Rhind, 1993, p. 7). Despite this, the RMSE is the measure most frequently recommended in the literature (e.g. Harley, 1975; Maling, 1989; Thapa and Bossler, 1992) and is commonly used by national mapping agencies (e.g. the Ordnance Survey and Australian Survey). Also, as long as only random errors remain (i.e. if systematic error has been eliminated) then the RMSE is a reasonable statistic to use (Hugh Buchanan, 1993, personal communication).

For both countries, in order to measure feature displacement, the smaller-scale maps were compared with the larger 1 : 50 000 scale map. Therefore, the error that was relevant to this study was the one that caused differences between the positions of the features at the different scales. Any error already present in the 1 : 50 000 scale map which was carried over to the smaller scales (such as surveying errors) would not affect experiments to determine differences between the maps. This was more the case when the smaller scales were derived directly from the 1 : 50 000 scale map, as for the Portuguese maps. The British maps have a complex lineage (see section 3.1.1) but, for the sake of this study, it was assumed that the error for the 1 : 50 000 scale map for both countries was incorporated in the smaller-scale maps.

4.1.1 Positional accuracy of the Portuguese maps

First, it was necessary to investigate how the two pairs of Portuguese maps differed in registration relative to each other (1 : 50 000 with 1 : 200 000; 1 : 50 000 with 1 : 500 000). As described in section 3.4, after the elimination of systematic error, the mean random error of the three Portuguese maps was approximately ± 0.3 mm. This error represented the positional accuracy of the triangulation points in relation to their *true ground positions*. However, because the smaller-scale maps were compared with the 1 : 50 000 scale map, the best solution was to find the *difference in the position between the maps*, rather than in relation to the ground truth (Hugh Buchanan, 1993, personal communication). In order to do this, the coordinates of the triangulation points at the two smaller scales were subtracted from the coordinates of the same points on the 1 : 50 000 scale map. The mean displacement (in RMSE terms) of the common six triangulation points between the 1 : 50 000 and the 1 : 200 000 scale maps was found to be *± 0.309 mm*. In the case of the 1 : 500 000 scale map, the mean displacement (in RMSE terms) was found to be *± 0.243 mm*. This distance between the triangulation points of the different maps accounts for errors in registration of each smaller-scale map with the 1 : 50 000 one.

Besides the registration error, further errors are introduced by drawing and digitising, which particularly affect features other than the triangulation points. In national mapping agencies it is common to accept drawing error within a half line width, approximately equal to *± 0.1 mm*. This is supported by Maling (1989), who suggests values for drawing error ranging from

± 0.06 mm to ± 0.18 mm. In addition, the digitising accuracy of the Portuguese maps using VTRAK was found to be larger than ± 0.06 mm (see section 3.2). Because the digitising was carried out using colour printed paper maps, there was an extra source of error due to colour registration. Errors in colour registration are on average ± 0.2 mm (Maling, 1989).

The total accuracy of the Portuguese maps, excluding generalisation (E_{total}), could be estimated by combining all of the different sources of error:

$$E_{total-200k} = \pm(0.309^2 + 0.1^2 + 0.06^2 + 0.2^2)^{0.5} = \pm 0.386 \text{ mm} \qquad (4.2)$$

$$E_{total-500k} = \pm(0.243^2 + 0.1^2 + 0.06^2 + 0.2^2)^{0.5} = \pm 0.336 \text{ mm} \qquad (4.3)$$

The smaller value of the estimated error for the 1 : 500 000 scale map (contrary to what would be expected) was probably due to the lack of precision in the calculation of the registration error (because only two triangulation points were used, as they were the only ones common to both the 1 : 50 000 and the 1 : 500 000 scale maps). Because the error found for the 1 : 500 000 is probably an under-estimation and, because of the need to be conservative, both values were rounded up to 0.4 mm. This error value was used as a basis for evaluation of the displacement of the Portuguese features. It was considered that any error exceeding the *0.4 mm cut-off point* was *more likely* to be due to generalisation than to accidental random errors.

4.1.2 Positional accuracy of the British maps

The accuracy of the 1 : 250 000 digital data was carried out to the OS standard for the 1 : 250 000 scale mapping, and was partly dependent on the generalisation of the data (Ordnance Survey, 1990). Field accuracy testing of the 1 : 250 000 performed by the OS revealed a RMSE of ± 0.5 mm (the value quoted in the 1992 OS digital map data catalogue) which is equivalent to 125 m on the ground. This testing was done by comparing the differences in the positions of well-defined points (such as road intersections) on the 1 : 250 000 digital data with their locations on the largest-scale mapping available for different locations around the country. However, some of the points selected will have probably suffered generalisation (Geoff Johnson, OS, 1993, personal communication). Because this would lead to an exaggerated value for the cut-off point, it was decided to try to determine the amount of random error excluding generalisation (see below), following a procedure similar to that used for the Portuguese maps.

The accuracy of the 1 : 625 000 digital data was to the OS standard for mapping at that scale (Ordnance Survey, 1987). It was therefore strongly dependent on the generalisation of the data. The map used as the basis for the 1 : 625 000 digital data was the Routeplanner map, which in some cases was adjusted in order to improve absolute accuracy (Ordnance Survey, 1987). In the case of the 1 : 625 000 data, the 1992 OS digital map data catalogue claims that an accuracy value is not available (Ordnance Survey, 1992). The reason for this is that the 1 : 625 000 scale map is so generalised that it is unsafe to quote an accuracy (Fred Neil, OS, 1993, personal communication).

It was therefore necessary to try to calculate the positional error excluding generalisation for both the 1 : 250 000 and the 1 : 625 000 scale data. As with the Portuguese maps, it was necessary to find out how the registration of the smaller scales differed in relation to the 1 : 50 000 scale map. However, there were no triangulation points drawn on the 1 : 250 000 and the 1 : 625 000 scale maps, and so the method used for the Portuguese maps could not be used. An alternative was therefore to use a theoretical error value in plotting the control. According to Maling (1989), the average error due to plotting the control has a value of *0.165 mm*. This was considered a reasonable estimate by the OS (David Rhind, OS, 1993, personal communication).

The fair drawing accuracy of both the 1 : 250 000 and the 1 : 625 000 scale maps was, according to the OS standards, within half a line width (i.e. *0.1 mm*). The digitising accuracy of both of these maps was usually within 0.1 mm of the centre of the line, although some points could have deviated 0.2 mm away from the line centre (Geoff Johnson, OS, 1993, personal communication). The worst case of *0.2 mm* was therefore used. For the British maps, error due to colour registration did not need to be considered, as the maps were digitised from separate plates of stable materials. The combined error estimate was the same for both scales:

$$E_{total-250k} = \pm(0.165^2 + 0.1^2 + 0.2^2)^{0.5} = \pm 0.278 \text{ mm} \qquad (4.4)$$

$$E_{total-625k} = \pm(0.165^2 + 0.1^2 + 0.2^2)^{0.5} = \pm 0.278 \text{ mm} \qquad (4.5)$$

These values were rounded up to ± 0.3 mm. Comparing the calculated value of the 1 : 250 000 scale map with the ± 0.5 mm accuracy testing obtained by the OS, it was decided to use a midpoint value of 0.4 mm instead. This is the same value as the one obtained by the Australian Survey for their 1 : 250 000 topographic database (Australian Survey, 1992). For the sake of this study, the same value was used for the 1 : 625 000 scale map (as it is unlikely that the 1 : 625 000 positional errors are inferior to the errors of the 1 : 250 000 scale map). Therefore, displacement values greater than 0.4 mm on both the 1 : 250 000 and the 1 : 625 000 scale maps were assumed to be more likely to be due to generalisation.

In the case of the British maps, a further experiment was carried out to compare the 1 : 50 000 scale map with the 1 : 10 000 scale map (see section 4.3). Therefore, it was also necessary to determine (as an accuracy value was not readily available) the cut-off point between generalisation and other error for the 1 : 50 000 scale map. As for the other British maps, the error due to map registration was taken to be *0.165 mm* and the drawing accuracy was equal to *0.1 mm*. As for the digitising error, all line and point features of the 1 : 50 000 digital map were digitised such that, compared with the original document, the

RMSE error was no more than *0.2 mm* at the 1 : 50 000 scale map (Ordnance Survey, 1991). When these three sources of error were combined, the accuracy estimate of *±0.278 mm* was obtained. This value was rounded up to *± 0.3* mm. Therefore, when comparing the 1 : 50 000 with the 1 : 10 000 scale map in Appendix A, the value of *0.3 mm* is used as the cut-off point between generalisation and random error.

To summarize, for the British 1 : 50 000 scale map, the cut-off point was equal to 0.3 mm, and in the case of the two smaller scales of both countries the cut-off point was assumed to be 0.4 mm. Below the cut-off point, the error could be due either to random errors or to generalisation, and it would not be possible to dissociate the two. However, as the size of the positional error increases, there is an increased likelihood that this would be due to *conscious* generalisation. This is not to say that above this cut-off point *all* error is due to generalisation. Occasionally, it may be possible that some of the error above the cut-off point might be due not to generalisation but to *unconscious* large errors, such as accidental gross errors. This, however, would be rare, as it is unlikely that large errors would remain undetected and uncorrected during the map-making process. This also assumes that random errors larger than the cut-off point are very unlikely which, as stated earlier, is probably a safe assumption.

4.2 Measurement of the generalisation effects of different features

Two main aspects are of relevance when measuring generalisation effects of different geographical features: how the actual measurement of the generalisation effects were executed, and how the different features were identified. This section is therefore divided into two parts. The first describes the different measurements selected to quantify generalisation effects. The second part describes the range of problems in applying these measurements in practice. This includes the identification of the terminal points of the different features, and the difficulties inherent when applying the measurements to *manually* generalised lines (rather than lines generalised automatically).

4.2.1 Descriptors for generalised features

For this study, a series of measurements was made to evaluate the different aspects of the generalisation effects for data of different types. These measurements were especially important because they enabled quantitative evaluations of generalisation to be carried out. The measurements performed on each feature class fell into three main categories:

- Change in features' characteristics.
- Change in the quantity of features.
- Change in features' position.

In addition, two types of GIS map manipulations were carried out: a point-in-polygon operation and an overlay operation. These map manipulations were chosen because they are typical GIS operations. If the effects of generalisation were to influence the results of these procedures, then it is probable that other more sophisticated GIS analyses would also be affected.

The line characteristics measured were line length and line sinuosity. Line length for each of the features was measured automatically by the GIS. Every GIS can carry out spatial type measurements such as 'distance between points' and 'perimeter of polygons'. These are some of the most common operations used in a GIS. The second line characteristic measured was the sinuosity of each feature. The amount of sinuosity of a line was quantified by adding up the absolute angle of change between each pair of consecutive vectors of a line. Within a GIS, this meant *every* point used to represent the line, with the exception of the first and last point. Carstensen (1990) called this measurement the 'cumulative angularity' of a line (illustrated in Figure 4.1).

The results of angularity measurements were especially vulnerable to methodological flaws, such as tiny undetected spikes in the lines, compared with the other measurements used. For this reason, spurious spikes in the lines had been eliminated (see section 3.5). If a line was completely straight, then its cumulative angularity was zero. However, for a non-straight line, the cumulative angularity was related to both its 'wigglyness' and its length. In order for the sinuosity of each line to be comparable, it was necessary to standardise the sinuosity measurement. There were two options: divide the cumulative angularity either by the number of segments, or by the total length of each feature.

Authors such as Jasinski (1990) and Müller (1987) have chosen to standardise the angularity measurements of their automatically generalised lines by the number of segments. Müller (1987), for example, divided the cumulative

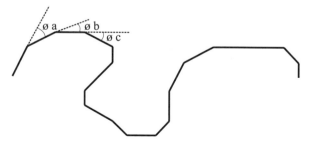

Figure 4.1 The measurement of the cumulative angularity of each line. The total angularity of each line is measured as the sum of all of the angles *a, b, c* and so on.

angularity of each line by the number of inter-segment angles (excluding the first and last point) using the following formula:

$$Ang = \left[\sum_{i=1}^{n-2} (1 + (\cos \phi + 1) \times 2)\right]\Big/(n-2) \tag{4.6}$$

where *Ang* is the average angularity of the line, ϕ is the angularity of each individual line segment and n is the number of points used to represent the line. However, this formula works most effectively with automatically generalised lines, when the generalisation of the same line (resulting from the use of different algorithms or from the same algorithm with different tolerances) is being compared. When applied to the manually generalised lines of this study, with their enormous variation in the numbers of segments, the equation was unable adequately to discriminate between lines that *appeared* to have different sinuosity. This was also evident, and acknowledged by Müller (1987), with his own data. For example, the angularity measure of a coastline (using the above formula) had a sinuosity value of 1.1378, while a coastline that appeared to have many more complex turns and meanders had a very similar value of 1.1484. Also, the final angularity value of manually generalised lines was affected by the existence of extra points, which were irrelevant to the line's total length or angularity. Because of these problems, all of the angularity measurements in this study were standardised using the length of the feature and were expressed in units of degrees per kilometre (rather than number of segments).

The second category of generalisation measurements was the change in the quantity of features. The initial intention was to measure the straightforward elimination of features (e.g. the removal of minor roads at small scales), but this was problematic. Large numbers of very small interrupted sections of roads and waterways were present at the large scales of the British maps. If each of those sections was counted as one feature, then a hugely exaggerated number of feature eliminations would have resulted. Therefore, feature elimination was calculated by measuring the total *length* of all of the same types of features present at different scales.

The third category of generalisation measurement was feature displacement. As scale reduces, and the available map space becomes smaller, entire features or parts of features might need to be displaced, in order to make room for other features or to make sure that features do not coalesce. This is *deliberate* displacement. The cartographer is sacrificing absolute accuracy (the position of the features relative to a coordinate system) in order to maintain or optimise relative accuracy (the position of the features relative to each other). In this study, this type of displacement is termed *global displacement*, as it affects a large portion of a feature or even, in certain circumstances, the whole feature. In other words, the feature on a particular map sheet might be completely displaced in order to avoid other features. Displacement can also occur as an indirect consequence of the simplification of lines and areas: as shapes become altered and lines straightened out, small sections of the features (sometimes as small as sliver polygons) will be displaced. In this study, this type of displacement is termed *local displacement*. With local displacement, particular sections of a line might lose detail (e.g. a particular road bend) when simplified and, as a consequence, local sections of a feature are displaced sideways.

In order to measure the sideways shift of the features caused by generalisation, the features from the smaller-scale maps were enlarged and overlaid with the equivalent features represented on the largest-scale map (1 : 50 000 for both countries). McMaster (1986) described two descriptors for measuring the sideways shift of automatically generalised lines: vector and areal displacement. However, for manually generalised lines, a key problem with the use of vector displacement techniques is the determination of a sampling rate. In order automatically to calculate the mean sideways shift of a manually generalised line, it is necessary to have a specified interval size. At each of these sample points, the distance between the two lines is measured. For the same line, one sampling rate might be appropriate for certain portions of the line, while for other parts of the line it could 'miss' the maximum displacement and therefore underestimate the displacement error. This is especially true with polygons generated by manually generalised lines, as these are very irregular, and the sampling rate that would suit one feature might well be unsuitable for another feature.

Four displacement measurements were carried out in this study: total areal displacement (the area between the original line and the line at the smaller scale); the number of polygons generated; the maximum displacement observed in each feature class; and the displacement of well-defined points (such as road intersections). The first two values were divided by the length of each feature (measured at the largest scale) so that the values for the different feature types and for the two countries were standardised and hence comparable. The formulas used in each of these cases were the following:

$$A_f/L_f \ (\text{m}^2/\text{m}) \tag{4.7}$$

in which A_f was the total areal displacement of a certain feature class f (in square metres) and L_f was the total length of features of the same feature class (in metres), and

$$P_f/L_f \ (\text{polygons/km}) \tag{4.8}$$

in which P_f was the total number of polygons of a certain feature class f and L_f was the total length of features of the same feature class (in kilometres).

The areal displacement was used as the *mean* indicator of the sideways shift of a line. It was calculated by overlaying the same line at two different scales and measuring the areas of the polygons generated by the non-matching overlap. To complement this indicator, the *maximum* displacement vector of each line and the displacement of *intersection* points were also measured interactively, using LITES2 editing software. While the maximum vector displacement pin-points the sections of the lines that show the largest displacement, the displacement of the intersection points allows the displacement of well-defined points (i.e. points that are known to be the same across the different scales) to be evaluated.

As well as these selected measures of displacement, others were also considered. Müller (1987), for example, described another measurement of displacement which he termed the Standardized Measure of Displacement (SMD):

$$\text{SMD (\%)} = 100(1 - (W - O)/W) \tag{4.9}$$

where W is the displacement in the worst case (the displacement between the actual line and a straight line linking the first and last points) and O is the observed displacement.

This last measure was rejected, as it was found to be unsuitable for the features in this study. The SMD is most appropriate for comparing the displacements caused by different simplification algorithms *of the same feature*. For automated methods, the maximum possible displacement value will be 100 per cent because, by definition, the first and last points will remain the same. However, with manual generalisation the whole feature can be displaced sideways and therefore the SMD can have values higher than 100 per cent. Because the worst case varied so much for manually generalised features, this measure was not appropriate for comparing the displacement of the different features.

Most of the above processing was carried out using Laser-Scan software and specially written code (where possible in the Laser-Scan macro language). The results of the measurements were written into report files (i.e. ASCII files). Most of the measurements were carried out on one map at a time. In the case of the displacement measures, however, all digital maps for each region were overlaid in register on a pair-wise basis, and the differences amongst maps observed. This entailed structuring Laser-Scan files and creating polygons, in order for the areal displacement to be determined.

All of the measurements described were applied to four different types of features – roads, boundaries, railways and rivers. These feature types were chosen because they were common across the maps of both countries. In order to compare the generalisation of the different maps, the individual features selected had, in most cases, to be shared by all of the map scales of each study area: this meant, for example, that many of the minor roads on the large-scale maps were discounted. The exceptions to this were the measurement of feature elimination and one of the overlay operations that took account of all of the features present. Because feature-against-feature comparisons were needed, it was necessary to indicate the limits of the features to be compared. The next section describes the problems encountered in implementing the generalisation measurements and in identifying the beginning and end of the features.

4.2.2 Problems in measuring different features across map scales

The measurement of the generalisation effects was not always straightforward. Even if the features were present at all three different scales, their classification within the maps could change. For example, some roads on the large-scale Portuguese map were classified as minor roads, but they had been re-defined as major roads on the small-scale map. In certain circumstances, this change in classification could be explained by differences in the dates of the maps (e.g. four years later the roads had been improved and upgraded), but this was not always the case. The misclassification meant that differences in generalisation effects between subclasses of feature could not be carried out. There was also a further problem of aggregation and disaggregation of features: for example, the outline of Canterbury at the largest scale becomes several distinct features at the scale of 1 : 250 000. It was therefore difficult uniquely to define the urban outline of Canterbury and hence measure its area.

Another important difficulty in the measurement of the generalisation was the identification of the limits of the features across map scales. Where alternative definitions exist, the definition of a feature to be measured can influence the results. Maling (1989) identified this problem as cartometry's most difficult task. It is difficult to determine where a feature starts and ends in order for the same points to correspond across different map scales. The two most straightforward ways of cutting features use either the edge of the study area or intersections with other features. However, both of these proved to be unsatisfactory and a third method was adopted.

Intersections are often used for the identification of the precise location of a geographical point (e.g. for the correction of aerial photographs or remote sensing images). However, if features were cut on the basis of intersections, this would have meant that the length of each feature would have been dependent on its own generalisation and the displacement suffered by the other feature. In other words, the measured length would be dependent on the *generalisation of two features*. Only in the case of some of the roads that start and/or end within the study area (e.g. the case of the British roads that finish when they intersect the Canterbury by-pass), did the intersection point with other features have to be used.

The other immediate way of cutting features is based on the edge of the study area. However, depending on the angle at which the features crossed the line delimiting the study area, length could change considerably (see Figure 4.2b). A small sideways displacement of one of the features at one of the scales could, in principle, lead to a considerable length difference between the features across the scales, not directly caused by generalisation. Cutting the features using the edge of the study area is only effective if the features at both scales 'leave' the study area at a similar angle and at the same point of the feature. In an hypothetical situation in which one feature intersects the edge of the study area at right angles and the same feature, at a different scale, runs

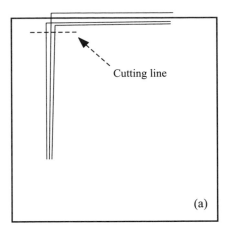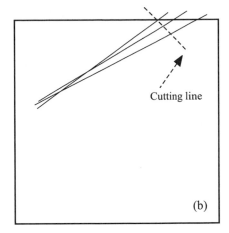

Figure 4.2 Two examples in which the edge of the study area would cut the features in an unacceptable way. A cutting line 'tailor made' for the individual features had to be used instead: (a) a small displacement of one of the features at one of the scales might mean that the features would have a very large difference in length; (b) features can have different lengths according to the angle at which the different features leave the edge of the study area.

parallel to the edge inside the study area, the length difference between them would be very large (see Figure 4.2a). To overcome this, most features were cut at a point approximately perpendicular *to the ends of the lines* at the three different scales, as illustrated in Figure 4.2.

There were further complications due to the way in which lines are generalised manually rather than automatically. When using an automated process, such as the Douglas–Peucker algorithm, to generalise a line, the points of the line that are considered most important (e.g. sharp bends) will be retained, and the ones found less essential in preserving the line's general shape will be eliminated. As a consequence, many of the same points are retained between the initial and the final automated generalised line (see Figure 4.3) and this makes it easier to apply certain descriptors (such as vector displacement, as defined by McMaster, 1987).

The displacement vectors of automatically simplified lines are always perpendicular to the generalised line segments. However, when a line resulting from manual generalisation was compared with the original line, it was difficult to determine which point on the line was the same across the two scales. It was therefore more difficult to define vector displacement. This was especially the case when the manually generalised line had been displaced along its entire length. On the whole, the measurements of generalisation effects were easier to carry out with features that had been generalised automatically rather than manually.

Manual generalisation   Douglas-Peucker algorithm

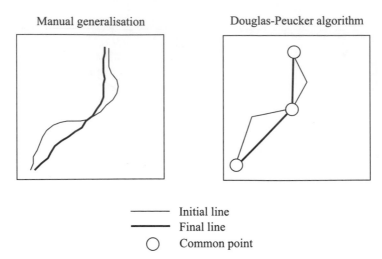

—————— Initial line
—————— Final line
○ Common point

Figure 4.3 The differences between the generalisation of a line using a manual process and using the Douglas–Peucker algorithm.

4.3 Validation of the use of the 1 : 50 000 scale OS map as the base map

In this study it is assumed that the 1 : 50 000 scale OS map was less generalised than the 1 : 250 000 and 1 : 625 000 scale maps. As will be seen in the results, this assumption is fair. However, it was felt important as a check to compare the 1 : 10 000 scale maps with the 1 : 50 000 scale map for any important generalisation effects. This was done in order to validate the use of the 1 : 50 000 scale OS map as the map against which the generalisation effects of the smaller-scale maps were measured. The results of this comparison, in terms of changes in line length and lateral displacement of features, are presented in Appendix A.

For this experiment, the 1 : 10 000 scale map was preferred to the 1 : 1250 and 1 : 2500 scale maps, for two reasons. First, the 1 : 50 000 is derived at least partly from the 1 : 10 000 scale map. Second, by using a 1 : 10 000 scale sheet (rather than an even larger scale map) more and longer features of the 1 : 50 000 scale map could be analysed. Only eight features were digitised from the 1 : 10 000 scale map for comparison with the same features of the 1 : 50 000 scale map. These were the only features that were common across the four different scales. As a consequence, the generalisation that the 1 : 10 000 scale map might have suffered in terms of elimination of minor features (as described by Harley, 1975) did not affect the features digitised. Many of these features (i.e. the roads and the river) were represented by a double line at 1 : 10 000, but only the centreline was digitised. Therefore the features digitised were not affected by the exaggeration of the true width of features that may have occurred.

Except for one of the features, all of the generalisation effects were found to be minimal (see Appendix A). This is confirmed by an unpublished experiment by the OS, in which 1 : 50 000 scale maps were compared with the OSCAR data (a digital road centreline network derived from OS maps at source scales of 1 : 1250, 1 : 2500 and 1 : 10 000) and it was found that there was a very good fit between the two sets of data (David Rhind, OS, 1993, personal communication). The 1 : 50 000 scale map can therefore be considered to be an accurate representation of the 1 : 10 000 scale map, although there are local-ised generalisation effects.

These results reinforced the decision to use the 1 : 50 000 scale OS map as the baseline from which the generalisation effects of the smaller-scale maps were determined. This is supported by the fact that the 1 : 250 000 and 1 : 625 000 scale maps are generalised versions of the 1 : 50 000 scale map (at least indirectly via the 1 : 63 360 scale map). Because for the Portuguese study area the 1 : 50 000 scale is the basic scale map – that is, it is the largest scale available from the IPCC – this comparison was neither possible nor necessary for the Portuguese maps.

4.4   Determining the fractal surface of each study area

An essential element of all landscapes is the roughness of the terrain. Physical features such as rivers adapt to the characteristics of the topography of a region, affecting the intrinsic characteristics of the lines. The more rugged the terrain is, the more features such as roads will be winding. This will have an important effect on the generalisation of such features. The most powerful way of measuring roughness of terrain is by determining the fractal surface of the study areas. Therefore, the fractal dimension of the topographic surfaces was calculated in order to control the effects of terrain on the analysis of gener-alisation effects.

The calculation of the fractal surface was based on contours from the 1 : 50 000 scale maps of both countries. The contours are equivalent to taking 'slices' of topography at a constant height. The longest contours that covered a large part of each study area were selected. The method used for measuring the fractal surface was the walking dividers method. Other methods for mea-suring fractal surfaces include the variogram and cell counting methods. They use the elevation values directly, while the dividers method uses the contour lines.

The dividers method has been the most common technique to determine the fractal dimensions of linear phenomena, although for fractal surfaces some researchers have pointed out that in certain circumstances it might underesti-mate the complexity of the topographic surface (Xia et al., 1991a). The fractal dimension is obtained from the slope of the linear regression of the log of the dividers spacing against the log of the measured length of the line. The slope of this regression is equal to $1 - D$, in which D is the fractal dimension of the curve. The fractal dimension of the *surface* can be obtained by adding *one* to the value obtained for the dimension of the contour lines by the dividers method (Klinkenberg and Goodchild, 1992). As well as calculating D, r^2 is also calculated in order to indicate how well the data points fit the regression line.

The three most important variables that need to be decided before running the walking dividers program are the initial step size, the final step length used and how the remainder is dealt with. Almost all 'walks' along a line result in a non-integer number of dividers being required to cover the line completely, and so it is necessary to devise a way of handling the remainder. The initial step size used was half the mean segment length size (i.e. the mean distance between consecutive pairs of points), as recommended by Shelberg *et al.* (1982). The final step length used was half the total length of the line. The remainder was treated using the approach most commonly taken, which is to add the remainder as a proportion of a divider. Aviles *et al.* (1987) found that this approach can produce slightly greater scatter in the plots and slightly higher values for the dimension. Care was therefore taken in the acceptance of plots with the least scatter ($r^2 \geqslant 0.95$). Klinkenberg and Goodchild (1992) followed a similar approach, in which they discarded contours if the scatter of points appeared too great or if the r^2 value fell below 0.90. In addition, all of the walks that generated a remainder greater than 2 per cent of the length of the line were discarded.

After the dividers are walked along the curve with one step size, the dividers are set to a larger step size. All step sizes were evenly distributed in log space. This eliminates biasing introduced when using linear regression to find the line of best fit and the slope of a log–log plot (Xia *et al.*, 1991a). The determination of each step size (S) was calculated by the following formula:

$$S = \log^{-1}\left(\frac{\log S_{max} - \log S_{min}}{n * 4^{-1}}\right)$$

$$\text{unless } n * 4^{-1} > \text{maximum number of walks} \quad (4.10)$$

in which S_{max} is the maximum step size (which for this study was set to half the total length of the line), S_{min} is the minimum step size (which for this study was equal to half the mean cord length), and n is the number of points of the line. The maximum number of walks (i.e. the number of times the dividers with different step sizes are walked along the curve) was set to 30. The critical number of walks was the minimum number of walks required for the log–log plot. The minimum value recommended ranges between five and eight for the number of walks (Xia *et al.*, 1991a). In none of the plots was there less than eight data points.

The fractal dimension was calculated by averaging the value of D for all the contours in which the scatter plots had a value of $r^2 \geqslant 0.95$. If the initial step size was found to be too small (causing a curvature in the scatter plot and a low value of r^2), the first values were discarded, as recommended by Xia et al. (1991a). In the case of the Portuguese study area, there were eight contours in which the final scatter plot had a value of $r^2 \geqslant 0.95$, and in the case of the British study area there were seven contours.

The fractal dimension of the Portuguese topography was found to be 2.31 (standard error of the mean = 0.016) and the fractal dimension of the terrain of the British study area was 2.18 (standard error of the mean = 0.016). According to Xia et al. (1991b), a D value of 2.2 represents the lower end of natural surfaces, while the highest fractal dimension found in their study was 2.83. They also found that the average fractal dimension of all the 191 different terrain surfaces studied gave a D value of 2.5, which the authors considered to represent the mean condition of land-surface topography. The low value of D found for both the British and Portuguese areas for this study implies that neither site was particularly irregular. Because both study areas had terrain that was not extremely irregular or hilly, then any variation between the lines at different scales was unlikely to be caused by the high-frequency oscillations of the contours. It was therefore possible to group features by feature type (e.g. roads) within each study area. If the terrain had been much more variable in roughness, then it might have been more appropriate to analyse the features individually.

4.5 Measurement of the effects of generalisation on GIS map manipulations

GIS map manipulations will normally involve the use of more than one feature type. The results of these analyses will reflect, therefore, a *combined* generalisation effect due to the generalisation of separate features. This is in contrast to more straightforward measurements (e.g. line length), in which generalisation influences only an individual feature. In order to determine how generalisation affected GIS map manipulation operations, a point-in-polygon and an overlay operation were carried out. In a vector-based GIS, a point-in-polygon operation identifies points contained within a specific area, while an overlay operation is used to bring together data at a common scale, using various Boolean operations. The processing of both of these GIS operations was done using ARC/INFO commands. For this, the Laser-Scan files had to be converted to ARC/INFO coverages.

Both the point-in-polygon and the overlay operations were repeated for the two countries and for each source scale. The original aim was to carry out the analysis using only the features that were common across the different map scales. For example, minor roads that were not represented on the small-scale maps were excluded from the analysis. This allowed the effects of the simplification and displacement of features on the analysis to be compared. Although this afforded greater comparability, it was felt that this did not represent what most GIS users would do in a real situation. The true impact of the generalisation process in GIS map manipulation operations is only felt when the elimination of features is taken into account as well. Therefore, in the case of the overlay operation, the analysis was repeated using *all* of the features, irrespective of whether the features were present across the different map scales.

A point-in-polygon operation is a spatial search function that identifies points contained within a polygon. Most GIS will perform various kinds of search function. The three basic parameters to be defined for such an operation are the targets to be found, the search area and the function that is to be applied (e.g. the total number or the average of the values within the search area). The point-in-polygon operation performed aims to be representative of the type used for environmental impact studies. A study by Cross (1989), for example, used a point-in-polygon operation to test the hypothesis that electromagnetic radiation from high-voltage transmission lines might increase the prevalence of childhood leukaemia. In her analysis, the points were the occurrences of childhood cancers and the search areas were buffers around the high-voltage transmission lines.

For the author's research, the same set of randomly located points was overlaid on different generalised areas to see how many points changed their allocation to the target polygons. Randomly located points were used in order to determine how generalisation of the search areas – rather than the change in location of the targets to be found – would affect the point-in-polygon analysis. The search areas used were road buffers derived from the roads of both countries from the three different scales (see section 6.1). It was necessary to use random points (rather than, for example, an equally spaced grid of points) so that the location of each point was independent. Points that were clustered (i.e. dependent) would have distorted the results of the analysis. A random point generator was used to generate both the x and the y coordinate value of each point. As the random point generator only generated numbers ranging from 0 to 99, it was then necessary to convert these numbers into suitable coordinates for the study areas of both countries.

An overlay operation was carried out to find an hypothetical site for a new countryside park within each study area. This type of overlay analysis, also sometimes called 'sieve mapping' or 'constraint mapping', is commonly carried out when using GIS for environmental management studies. Vaughan (1991), for example, reported how an overlay operation using GIS could be useful for locating a waste disposal facility within a region. It is also one of the GIS applications that is often considered to be advantageous over traditional manual methods. Manual overlay operations, such as the McHarg overlay method (McHarg, 1969), involve the manual handling of transparent overlays.

Overlay analysis using a GIS is more precise, less labour intensive, there is no restriction of the number of overlays that can be used, and different computations can be easily and quickly made (e.g. measuring the area of a site for the new waste disposal facility).

For the two GIS map manipulations, a different set of features was used. The point-in-polygon procedure only used road data in order to investigate a single feature type. For the overlay operation, the maximum possible number of feature types was used. This included the main line features (roads, rivers and railway lines), plus the position of the railway stations and the outline of the urban areas. More than one feature type was chosen in order to evaluate the *combined* generalisation effect as a result of the generalisation of separate features. The overlay operation was carried out first using only the features that were common across the different map scales, but it was then repeated using *all* of the features. This was done in order to simulate the *full* impact of the generalisation process that would be typical of a GIS map manipulation operation (i.e. taking into account the elimination of features as well). The next section describes how selected lines were generalised using an automated generalisation procedure in order to compare manual generalisation effects with automated ones.

4.6 Quantification of the effects created by the Douglas–Peucker algorithm

In order to compare the generalisation effects resulting from manual generalisation with the effects resulting from line simplification by use of mathematical algorithms, selected lines from the larger-scale maps were generalised using the Douglas–Peucker algorithm. One of the reasons for choosing this particular algorithm for this study was because it is one of the most commonly available algorithms in GIS used to simplify lines. The Douglas–Peucker algorithm 'has become one of the most favourite line simplification algorithms in use today' (Weibel, 1986, p. 25). Another reason for its selection was that several independent authors have found that it causes fewer unwanted generalisation effects when compared with other line simplification algorithms. McMaster (1987) has shown, for example, that the Douglas–Peucker algorithm causes less overall distortion (e.g. less areal displacement) than other line simplification algorithms. White (1985) tested a series of line simplification algorithms in order to determine which retained more critical points (the points considered to be the most important for maintaining the line structure) and found that the Douglas–Peucker algorithm was the most effective. Müller (1987) also found that, of the seven algorithms tested, the Douglas–Peucker algorithm was the best at preserving the degree of complexity of the lines in terms of their fractal dimension.

Like most line simplification algorithms, the Douglas–Peucker algorithm selects critical points on the basis of the relationships of points and their neighbours (McMaster, 1987). However, unlike other algorithms, it uses a global routine because it processes the entire line rather than only sequential sections of it (Cromley, 1992a). Crucial to the operation of the algorithm is the setting of the *generalisation tolerance*. It is the size of this tolerance that determines how many points of the line will be eliminated (see Figure 4.4). When the algorithm is first applied, the first endpoint of the line is set as an initial *anchor point* and the other endpoint of the line is designated as the current *floating point* (see Figure 4.4a). The following description (Cromley, 1992a, p. 219) explains what happens next:

> If the perpendicular distance between each intervening point and the straight line connecting the anchor and floating points is less than the [generalisation] tolerance, all of the intervening points are deleted and the anchor point is updated to the floating point. Otherwise, the intervening point having the maximum perpendicular distance becomes the new floating point and the intervening point check is repeated. These iterations continue until the anchor point becomes the last endpoint of the line.

In Figure 4.4b a line is compared before (the full line) and after (the dashed line) generalisation by the Douglas–Peucker algorithm. It can be seen that the line has lost 11 points and has retained 10 points (including the first and last point, which are never affected by the algorithm). If the generalisation tolerance had been larger, the line could have lost more points. In an extreme case (i.e. for a very large tolerance), the line is left with only the first and last points.

The main problems that arise from the use of the Douglas–Peucker algorithm are related to the creation of lines that are excessively spiky (Visvalingam and Whyatt, 1990). These can (in certain circumstances) lead to more serious topological errors, as identified by Müller (1990b), such as lines crossing back on themselves in very complex regions (e.g. in estuaries). However, these problems only occur when very large tolerances are used. Douglas and Peucker (1973) themselves recognised this problem and termed the use of their algorithm for very large tolerances as the production of 'line caricatures'. This has led authors to recognise that the Douglas–Peucker algorithm can only be successfully applied for modest scale reduction (Visvalingam and Whyatt, 1990). For this study, the tolerances used were very small, ranging from 0.0002 mm to 0.1 mm (see Appendix B). Therefore, these potential problems did not arise.

For this experiment, one river, railway, road and boundary were selected from the 1 : 50 000 scale map of each country and were generalised using the Douglas–Peucker algorithm. As much as possible, all of the features selected crossed the border of the study area at both ends. None of the features (with the exception of the British boundary) therefore had its beginning or end determined by an intersection with another feature. This was in order to avoid

(a)

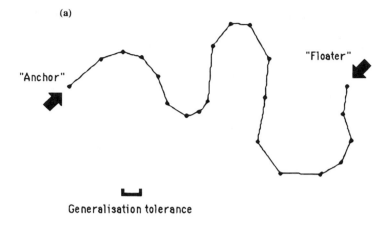

(b)

Figure 4.4 Line simplification carried out by the Douglas–Peucker algorithm, as illustrated by the GISTutor published by GeoInformation International (Raper and Green, 1989): (a) before the simplification; (b) after the simplification – the simplified line is compared with the original line.

the length of the feature being influenced by the displacement of another feature. The features selected in the Portuguese case were the railway and the county boundary, because they were the only features available for those particular feature classes. The Ribeira Grande was the main river present in the study area and the road 245 (rd ac 1) was the road with the most detail. In the British case, the Great Stour river was selected, as it was the only river present across all three scales. The railway to Ramsgate (rail ac 2) was chosen as it suffered the least change in length. The A251 road was the longest road with the most detail, while also beginning and ending outside the limits of the study area. Finally, the boundary chosen was the one that suffered the least displacement.

When performing the line generalisation using the Douglas–Peucker algorithm, the number of points used to represent the lines was reduced. The *automated* reduction needed was determined by the number of points required to represent the same lines on the smaller-scale maps derived from *manual* generalisation. In other words, the number of points on the lines resulting from the automated and the manual processes were equivalent, although the shape of the lines could be different. The assumption here is that the manual cartographer is able to carry out the generalisation efficiently – at least in a graphic sense – and that the machine is matched against this human filtering process. The initial number of points used to represent each feature and the Douglas–Peucker generalisation tolerance needed to achieve the required final number of points in each case are listed in Appendix B.

The reason for selecting only one feature of each type (one road, one river, one railway and one boundary) for both countries was that the automatic generalisation of lines was a time-consuming process. The process entailed iterative trial and error, in which the same line was generalised repeatedly until the right number of points was reached. If, after generalising, the line using the Douglas–Peucker algorithm had a number of points less than the line generalised manually, then the 1 : 50 000 scale line was generalised again using a smaller tolerance. If, on the contrary, the number of points was still greater than the manually digitised line, then more points needed to be eliminated and the 1 : 50 000 scale line was generalised using a larger tolerance.

The first step of the automated generalisation of the lines was the elimination of spurious points. When the manually generalised lines were converted to a digital form, spurious points could have been introduced. These are collinear points which, if removed, do not affect the length and angularity of the digital lines. Because in this experiment the number of points was used as the indicator of the amount of automated generalisation needed, it was necessary to eliminate them. For example, take a hypothetical situation in which a river at the largest scale has 200 points and the same river at the smallest scale requires 150 points. If the river at the largest scale has 50 *spurious* points, then the automated generalisation of the 1 : 50 000 scale river will only eliminate these particular points and, in effect, the line remains unaltered. Therefore, the features used in this experiment had to be filtered beforehand using the Douglas–Peucker algorithm until the 'critical' number of points was reached. This number was defined as the minimum number of points needed to represent the digital line. To obtain the critical number of points, an iterative process was required in which the lines resulting from the filtering were measured in terms of length and angularity. If the length or angularity had changed, then the tolerance value had been too high and filtering was repeated with a smaller tolerance. Finally, when all selected lines had been generalised using the Douglas–Peucker algorithm, the results of the effects of generalisation caused by this algorithm were compared with the effects of manual generalisation of the same lines.

The measurements of map differences performed in this study not only enabled the generalisation effects at different scales to be quantified, but they also served as the basis for testing and improving the quantitative techniques to be used to measure generalisation effects in the future. By actually applying the descriptors to a sample of map sheets, a better assessment can be made of their practicability and effectiveness. The next two chapters describe the results of measuring the generalisation effects for the study areas.

CHAPTER FIVE

The effects of generalisation on individual features

In order to measure how generalisation would affect the use of a GIS, a series of GIS operations was performed, on both manually generalised and automatically generalised features. The *manually* generalised maps were obtained by digitising the manually drawn paper maps from different source scales. The *automatically* generalised features were obtained by generalising features from the largest-scale digitised maps for Canterbury and Fronteira, using the Douglas–Peucker algorithm. In this chapter and the next, I present the results of the GIS operations performed on these maps, starting with relatively simple operations and progressing to increasingly complex map manipulations. Changes in line length and line sinuosity, elimination of features and feature displacement caused by generalisation are described in this chapter, and in chapter 6 I evaluate how generalisation alters the results of GIS-based analysis.

The first type of operation described in this chapter is the measurement of line length and line sinuosity. Both of these characteristics are fundamental in terms of the description of the individual feature types and in terms of analysing the impact of generalisation on simple GIS map manipulations. This is followed by an investigation of feature elimination. Next, the measurement of the sideways shift of the different line features is presented. This proved to be a crucial measurement to help to interpret the impact of generalisation. In order to measure the magnitude of the sideways shift of the different line features thoroughly, two main measurements were necessary: the maximum vector displacement and the mean areal displacement of a line. All of these measurements, with the exception of feature elimination, were carried out for both manually generalised lines and the automatically generalised lines.

With a few exceptions, most of the GIS operations were performed on four types of features – roads, boundaries, railways and rivers. The spatial distri-

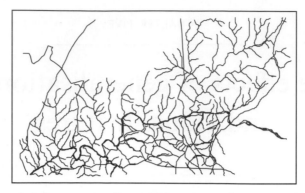

Figure 5.1 Fronteira's rivers, roads, railway and boundary at the source scale of 1 : 50 000. The Portuguese study area is approximately 16 km long and 10 km wide. Data taken from the IPCC 1 : 50 000 map sheet number 32-D (Sousel).

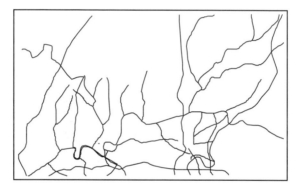

Figure 5.2 Fronteira's rivers, roads, railway and boundary at the source scale of 1 : 200 000. Data taken from the IPCC 1 : 200 000 map sheet number 6.

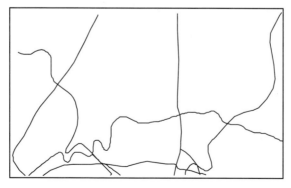

Figure 5.3 Fronteira's rivers, roads, railway and boundary at the source scale of 1 : 500 000. Data taken from the IPCC 1 : 500 000 map of Portugal.

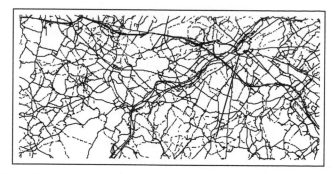

Figure 5.4 Canterbury's rivers, roads, railways and boundaries at the source scale of 1 : 50 000. The British study area is 20 km long and 10 km wide. Data taken from the OS 1 : 50 000 Landranger map sheet numbers 178, 179 and 189. © Crown copyright.

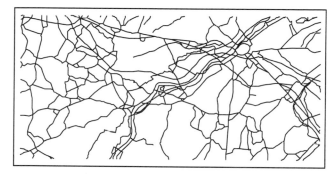

Figure 5.5 Canterbury's rivers, roads and railways at the source scale of 1 : 250 000. The district boundaries are not represented at this scale. Data taken from the OS 1 : 250 000 Routemaster map sheet number 9 (South-East England). © Crown copyright.

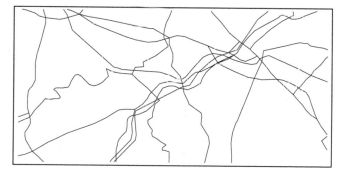

Figure 5.6 Canterbury's rivers, roads, railways and boundaries at the source scale of 1 : 625 000. Digital data taken from the following OS analogue maps: routeplanner, administrative and topographic. © Crown copyright.

bution of these features for the two study areas (Fronteira and Canterbury) for the different scales is shown in Figures 5.1–5.6. These feature types were chosen because they occur on all the maps, and therefore the same operations could be repeated for the maps of both countries for all of the scales.

To determine how the generalisation of the different scale maps affected the different measurements and GIS operations, the individual features used had, in most cases, to be shared by all of the map scales of each study area. For example, although there were 45 roads within the region of Fronteira at the largest scale, only four roads remained at the smallest scale. Therefore, it was these four roads, common to all the three scales, that were used for most of the measurements. This meant that, for the majority of the operations performed, many of the minor roads on the large-scale maps were discounted. Exceptions to this were the investigation of feature elimination, which took account of all the features present, and the case of the overlay operation (described in the next chapter) which, besides being carried out using only the common features, was repeated using all of the features.

5.1   Measurements of line length

Line length is one of the most basic characteristics of vector geometry measured by a GIS. Finding the distance between two railway stations, determining the length of a boundary that separates two countries, determining the total length of river within a watershed, calculating the length of roads built since last year, and so on, are all examples of GIS operations that depend on the accurate measurement of the length of a certain line feature. While some of these measurements might occasionally be carried out in the terrain (e.g. river length), most are more practically done using a map, and others (e.g. a boundary between two counties) can only exist as features on a map. Previous studies have analysed how scale affects the length of features that are measured in a map (see section 2.2.2). In this study more comprehensive measurements were carried out to investigate how changes in line characteristics influenced GIS map manipulations.

5.1.1   Length change on manually generalised maps

To determine the change in length of the different types of features for both the British and the Portuguese maps, all features common to the smallest-scale maps were used. In the case of the Portuguese maps, the following features that were present in all three maps were measured for the three scales: four roads with a total length of approximately 34 km; two rivers, approximately 24 km long in total (the Ribeira Grande and the Ribeira de Lupe); and one county boundary of 16 km and one railway line of 14 km (lengths as measured on the 1 : 50 000 scale map). The change in line length when the same features were measured at the different scales is shown in Figure 5.7. It can be seen in this figure that all feature types decreased in length as the scale of the maps was reduced.

When the features were generalised from the 1 : 50 000 scale map to the 1 : 200 000 scale map, two classes of length reductions were observed. Road and railway lengths had the smallest reduction, and they both had the same rate of percentage reduction. This small change in length was because at the largest scale these features were relatively straight. On the other hand, the rivers and the boundary each showed a similar rate of length change, but one that was larger than that for the roads and railway. When the features were further generalised on the 1 : 500 000 scale map, the roads, the rivers and the railway maintained a similar reduction rate, but the rate for the boundary was very different. At a scale of 1 : 500 000, the county boundary was the feature the length of which showed the greatest reduction, becoming only 77 per cent of its initial length. The boundary at the 1 : 500 000 scale, unlike that at the larger scales, no longer followed a section of the Ribeira Grande (the actual boundary) but cut straight across it. Along a section of the other river, the Ribeira de Lupe, where the boundary still followed the river, the river (and hence the boundary) was straightened out. Finally, along other sections of the

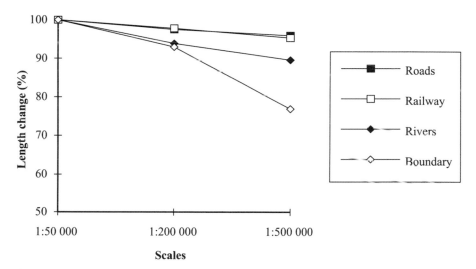

Figure 5.7 Inter-scale changes in the lengths of features common to all Portuguese maps. The absolute lengths of the different feature types at the scale of 1 : 50 000 are the following: roads (33 826 m); railway (14 154 m); rivers (23 831 m); and boundary (16 237 m).

boundary, the generalised boundary cut across several stretches of the original boundary and much of the detail was lost.

In the case of the British study area, the following features found in all three maps were measured for the three scales (see Figure 5.8): one river, approximately 18 km long (the Great Stour); two railways lines, with an approximate total length of 35 km; three district boundaries, with an approximate total length of 25 km; and 13 roads with a total length of approximately 85 km (lengths as measured on the 1 : 50 000 scale map).

Most British features showed a smaller linear decrease in length at the smaller scales compared with the Portuguese maps. The only feature type that did not decrease in length was the railways, which on average actually showed a slight *increase* in length (1.3 per cent). Between the largest and the smallest scale, the railway to Ramsgate had an increase in length of 118 m (0.8 per cent), while the railway to Dover had an increase in length of 353 m (1.8 per cent). This increase in length was due to an increase in the angularity of certain bends and the displacement suffered by the railways on the smaller-scale maps. Because railways tend to be very straight features, there was no potential for a decrease in length due to localised meanders being straightened out.

Figures 5.7 and 5.8 show the inter-scale changes in length for all the features aggregated by feature type, which is equivalent to their *average* change in length. However, when *individual* features were considered, the changes in some lengths were influenced by the generalisation of other features. For example, some roads contained within the map sheet had their start or end-

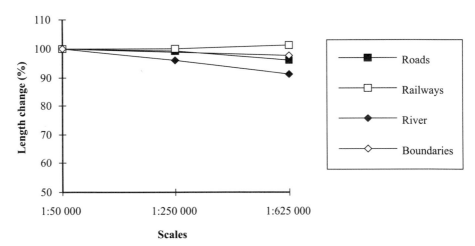

Figure 5.8   Inter-scale changes in the lengths of features common to all British maps. There is no value for the district boundary at the scale of 1 : 250 000, as it was not shown on the map. The absolute lengths of the different feature types at the scale of 1 : 50 000 are the following: roads (84 660 m); railways (34 739 m); river (17 755 m); and boundaries (25 059 m).

points established according to intersections with other roads. As roads were generalised at the smaller scales – in other words, simplified and displaced – so the start and endpoints of the roads were modified and the roads either became smaller or longer. The inter-scale changes in length for the four Portuguese roads are shown in Figure 5.9; and the inter-scale changes in length for the 13 British roads are shown in Figure 5.10.

It can be seen from both figures that individual roads had different patterns of length change as the scale of the map decreased. In the British case, for example, the road with code 23 had a large increase in length. Road 23 is the circular road that surrounds Canterbury, and was lengthened to make room for the exaggerated representation of the city at the scale of 1:625000. In contrast, the British slip road with code 17 has a sharp decrease in length, as it was 'crushed' between the enlarging circular road and the displacement northwards of the road with code 9 (the dual carriageway A2). It is therefore important to note that individual features showed greater variations in length with the change in scale than the mean values for that class of features would indicate. In the British case, the variation between roads increased progressively with scale. Intra-class variations can be considerable and this has important implications for the 'reverse engineering' of generalisation in GIS.

The results of this section indicate that changes were much more predictable in some features than in others. For example, rivers had a sharper decrease in length than roads, because in general they had higher sinuosity than roads, and when generalised these bends were removed (see section 5.2).

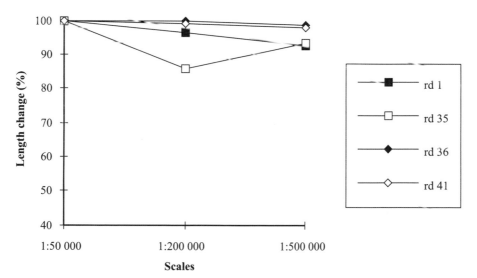

Figure 5.9 Inter-scale changes in the lengths of the four different roads of the Portuguese maps.

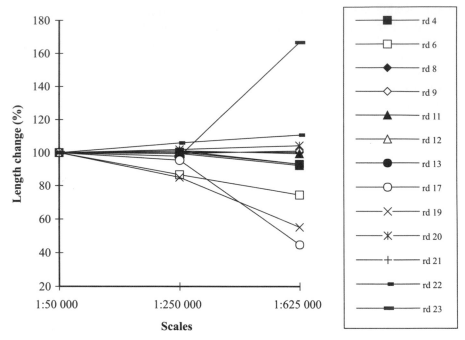

Figure 5.10   Inter-scale changes in the lengths of the 13 different roads of the British maps.

On average, no international differences were evident for roads and rivers, although the pattern of the percentage of length change for railways and boundaries was very different in both countries. The boundary in the Portuguese maps, for example, had a distinct length change, in which it suffered more simplification than any other feature of either country. It is possible that boundaries are considered less essential to the integrity of the map than more physical features.

It is commonly assumed that line features decrease in length as the map scale becomes smaller (as described in section 2.2.2), but in this study *occasionally* the reverse happened. As discussed above, this was the case with the circular road around Canterbury. The two British railway lines also increased in length, albeit slightly. One of the reasons for this counter-intuitive effect could have been that both of the railway lines had bends that were exaggerated – and in the case of the railway to Dover, the circular road pushed the railway southwards, lengthening it even more.

5.1.2   Length change for automatically generalised features

For this experiment, one river, railway, road and boundary were selected from the 1 : 50 000 scale map of each country and were generalised using the

Table 5.1 The lengths of the selected Portuguese features when generalised manually and automatically. The percentage change length (in brackets) is in relation to the original length at the scale of 1 : 50000 (in brackets below the feature name).

	Scale change	Railway (14 154 m)	Road (10 463 m)	River (21 879 m)	Boundary (16 237 m)
Manual generalisation	50k to 200k	13 846 m (98.8%)	10 071 m (96.2%)	20 582 m (94.1%)	15 125 m (93.1%)
	50k to 500k	13 518 m (95.5%)	9706 m (92.8%)	19 369 m (88.5%)	12 494 m (76.9%)
Automated generalisation	50k to 200k	14 132 m (99.8%)	10 457 m (99.9%)	21 829 m (99.8%)	16 210 m (99.8%)
	50k to 500k	14 090 m (99.5%)	10 384 m (99.2%)	21 827 m (99.8%)	16 113 m (99.2%)

Table 5.2 The lengths of the selected British features when generalised manually and automatically. The percentage change length (in brackets) is in relation to the original length at the scale of 1 : 50 000 (in brackets below the feature name).

	Scale change	Railway (14 886 m)	Road (10 718 m)	River (17 755 m)	Boundary (8410 m)
Manual generalisation	50k to 250k	14 853 m (99.8%)	10 756 m (100.4%)	17 051 m (96.0%)	—[a]
	50k to 625k	15 004 m (100.8%)	10 757 m (100.4%)	16 180 m (91.1%)	7920 m (94.2%)
Automated generalisation	50k to 250k	14 884 m (100.0%)	10 705 m (99.9%)	17 652 m (99.4%)	—[a]
	50k to 625k	14 851 m (99.8%)	10 660 m (99.5%)	17 342 m (97.7%)	8299 m (98.7%)

[a] There is no value for the district boundary at the scale of 1 : 250 000, as it was not represented on the digital map data.

Douglas–Peucker algorithm. As described in section 4.6, the number of points used to represent each particular line was exactly the same for the automated and the manually generalised versions of that line. In Tables 5.1 and 5.2 are shown the results of measurement of the length of each feature, derived from both manual and automated generalisation methods for both countries. All of the lines had the same beginning and endpoints and the variation of length was therefore due solely to the generalisation of the lines.

It can be seen from the tables that the length change for the lines gener-alised automatically was much lower than that for the equivalent lines gener-alised manually. For example, with manual generalisation the Portuguese river (the Ribeira Grande) at the scale of 1 : 500 000 was only 88.5 per cent of the length of the river at the 1 : 50 000 scale. But when the automated gener-alisation was carried out, between 1 : 50 000 and 1 : 500 000, the river was 99.8 per cent of its original length. At the original scale, the river needed 592 points to represent its length and shape (see Appendix B), but only 161 points were needed at 1 : 500 000 with automatic generalisation (the same number of points as for the 1 : 500 000 manually generalised map). For the same number of points, there was considerably less length change.

Similarly, in the British case (see Table 5.2), the Great Stour, measured at the scale of 1 : 625 000, used 42 points for its representation and had a length that was 91.1 per cent of that for the scale 1 : 50 000 (at which it was represent-ed by 441 points). But when the river at the scale of 1 : 50 000 was generalised using the Douglas–Peucker algorithm until the number of points was reduced from 441 to 42, the length was 97.7 per cent of the original. In general, one scalar property of lines (their length) was preserved better by automated than by manual generalisation – at least with the Douglas–Peucker algorithm.

5.2   Measurements of line sinuosity

The sinuosity of each feature was quantified by adding up all of the changes in angles of the line and dividing this cumulative angularity by the length of each feature (see section 4.2.1). The sinuosity values are presented in units of degrees per kilometre. If a line was completely straight, then its angularity was zero, but for a non-straight line its angularity per unit length (with spurious spikes eliminated) was proportional to its 'wigglyness'.

5.2.1   Sinuosity change on manually generalised maps

In order to measure the angularity of the different feature types, a mean value was calculated based on the value of angularity per kilometre of each individ-ual feature. For both countries, the rivers had the highest sinuosity and the human-made features (railways and roads) had the lowest (see Table 5.3). In

Table 5.3 Mean angularity per kilometre for all the Portuguese and British features.

| | Angularity (degrees/km) | | | |
	Roads	Railways	Rivers	Boundaries
Portugal				
1 : 50 000	62	72	284	171
1 : 200 000	42	50	114	110
1 : 500 000	20	42	81	57
Great Britain				
1 : 50 000	72	27	338	174
1 : 250 000	55	27	114	—[a]
1 : 625 000	20	19	44	82

[a] There is no value for the district boundary at the scale of 1 : 250 000, as it was not represented on the map.

general, both countries showed a similar angularity per feature type for the different scales. The feature type with the most striking difference between the two countries was the railway line. The Portuguese railway line had a much higher angularity than the British one. Along one section of the Portuguese railway line, there was a distinctive loop that contributed to the large angularity value.

For both countries, the angularity reduced sharply with smaller scales for almost all feature types. The exception to this was the British railway lines, which had the same mean angularity per kilometre at both the 1 : 50 000 and the 1 : 250 000 scales (see Table 5.3). This was due to one of the two British railway lines having more angularity per kilometre at the scale of 1 : 250 000 than at the largest scale, because some of the bends were made sharper.

The change in the angularity per kilometre for the Portuguese features is shown in Figure 5.11. It can be seen that all the Portuguese features lost sinuosity at the smaller scales. For the Portuguese railway and rivers, the rate of sinuosity reduction was greater between the 1 : 50 000 and the 1 : 200 000 scale maps than between the 1 : 200 000 and 1 : 500 000 scale maps. The roads and the boundary had a relatively steady reduction in sinuosity between the three scales.

For the British features, the change in the angularity per kilometre is shown in Figure 5.12. The British river had a very similar trend to the Portuguese rivers in the reduction of the angularity per kilometre with the reduction in scale. The highest reduction in angularity happened between the first two scales. The further reduction of the sinuosity of the river, which occurred between the 1 : 250 000 and the 1 : 625 000 scale maps, was lesser in extent. This indicates that for the river features most of the localised meanders were lost in the transition between the large- and the medium-scale maps (a loss of

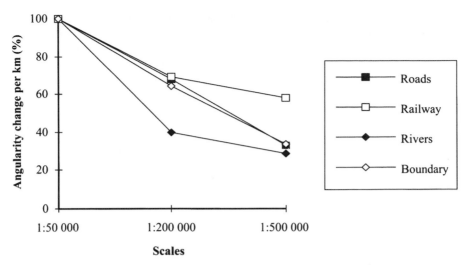

Figure 5.11 Inter-scale changes in the mean angularity per kilometre for features common to all Portuguese maps.

60–65 per cent of their angularity). This meant that for the next transition (between the medium- and the small-scale maps) there was not much angularity left in the rivers to be lost.

In the case of the British roads and railways, the trend in the reduction of sinuosity with scale was the opposite of what happened with the Portuguese

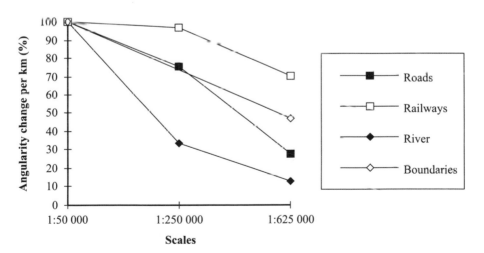

Figure 5.12 Inter-scale changes in the mean angularity per kilometre for features common to all British maps. There is no value for the district boundary at the scale of 1 : 250 000, as it was not shown on the map.

man-made features. Both the roads and the railways in the British region had a higher decrease in angularity between the last two scales (1 : 250 000 and 1 : 625 000 scale maps). In contrast, in the Portuguese case there was either a higher decrease in angularity between the first two scales (in the case of the railway) or there was a steady decrease between the scales (in the case of the roads).

The change in the angularity was related to the initial amount of sinuosity at the larger scale. The higher the initial sinuosity was, the more the feature's angularity was reduced at a smaller scale. Figure 5.13 shows the Pearson correlation (significant at the 0.05 level) between the mean angularity of each feature type at the largest scale (the 1 : 50 000 scale for both countries) with the change in angularity experienced by each of those feature types. This change in angularity was calculated by subtracting the mean angularity of each feature type at the smallest scale (1 : 625 000 scale for the British features and 1 : 500 000 scale for the Portuguese features) from the one at the largest scale. The correlations were still significant if the Portuguese and British features were treated separately.

By looking at one particular feature type in detail, it is possible to investigate the variability of angularity within feature types. In Figures 5.14 and 5.15 are shown the changes in angularity of individual roads for both countries. It can be seen that, as shown for the change in length, each individual road,

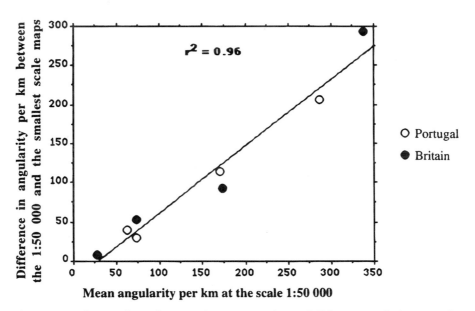

Figure 5.13   The correlation between the mean angularity of all features at the largest scale with the change in mean angularity suffered by each of those features between the largest and the smallest scales.

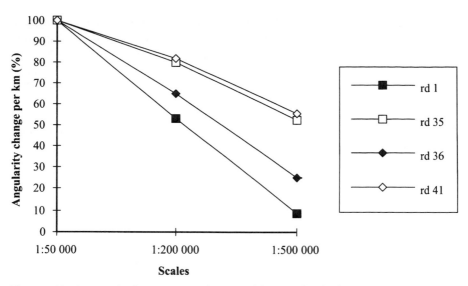

Figure 5.14 Inter-scale changes in angularity per kilometre for the four Portuguese roads.

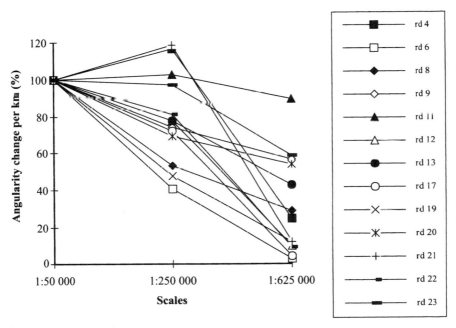

Figure 5.15 Inter-scale changes in angularity per kilometre for the 13 British roads.

irrespective of the country, had its own pattern of sinuosity change. For both countries, some roads had a steeper decrease in angularity per kilometre between the first two scales, other roads lost most of their sinuosity between the last two scales, and yet other roads had a steady decrease in angularity between the three scales. Three roads in the British maps had an *increase* in angularity per kilometre with the decrease in scale: road ac 4 (the B2068), road ac 11 (the M2 motorway), and road ac 21 (the A290). This increase is caused by gentle corners being replaced by more acute angles. The Portuguese roads showed much less variability.

5.2.2 Sinuosity change for automatically generalised features

In order to analyse the change in angularity for lines generalised automatically, the same lines as described in section 5.1.2 were used. The changes in angularity per kilometre for the selected road, railway, river and boundary, generalised both by manual and automated methods, are presented in Figures 5.16–5.19. In Figures 5.16 and 5.18 are shown the changes in angularity per kilometre for the lines generalised manually in the Portuguese and British cases respectively. In Figures 5.17 and 5.19 are shown, for the Portuguese and British study areas respectively, the change in angularity per kilometre for the same lines generalised by the Douglas–Peucker algorithm.

For both countries, the lines that resulted from generalisation by the Douglas–Peucker algorithm all had a smaller decrease in sinuosity per kilo-

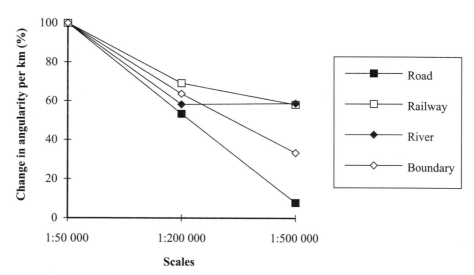

Figure 5.16 Inter-scale changes in angularity per kilometre for the *four* Portuguese manually generalised lines (cf. Figure 5.11, which shows the mean angularity for *all* features).

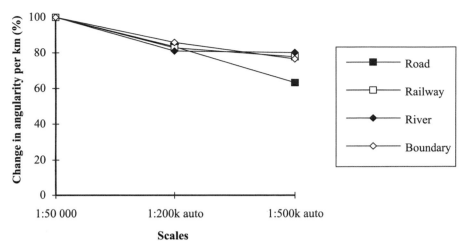

Figure 5.17 Inter-scale changes in angularity per kilometre for the four Portuguese automatically generalised lines.

metre than the lines generalised manually. This was especially evident in the Portuguese case. The Portuguese road 254 (road ac 1), for example, when generalised manually became almost a straight line at the scale of 1 : 500 000 (with only 7 per cent of the initial angularity left). This high percentage loss of angularity was due to the disappearance of a loop in the road at the smaller

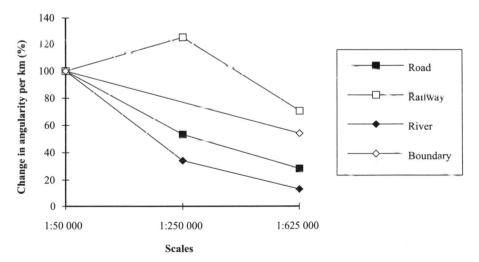

Figure 5.18 Inter-scale changes in angularity per kilometre for the *four* manually generalised lines, for Great Britain (cf. Figure 5.12, which shows the mean angularity for *all* features). There is no value for the district boundary at the scale of 1 : 250000, as it was not shown on the map.

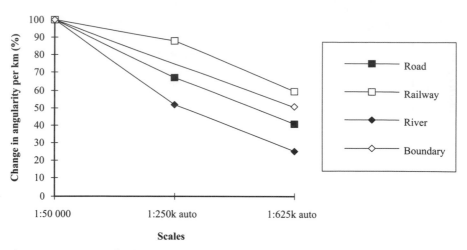

Figure 5.19   Inter-scale changes in angularity per kilometre for the four automatically generalised lines, for Great Britain. There is no value for the district boundary at the scale of 1 : 250 000, as it was not represented on the map.

scale. In contrast to this, when the road was generalised automatically the loop did not disappear and the road retained more than 50 per cent of its original angularity at the smallest scale.

For both countries, and with few exceptions, the lines generalised automatically retained a similar *pattern* of angularity change as the manually generalised lines. In the Portuguese case for example, the pattern of angularity change for the Ribeira Grande and road 254 was very similar in both cases (although the decrease in angularity was smaller in the automated case). In both the manual and automated generalisation, the river had a higher loss of angularity between the first two scales, while the road maintained a steady decrease in angularity.

As can be seen in Figures 5.16 and 5.17, the Portuguese road showed more of a decrease in angularity than the river, both when the lines were generalised manually and automatically. This was due to the amount of point loss in each case. The degree of generalisation caused by the Douglas–Peucker algorithm depends on the tolerance used. In this study, the tolerance value was determined by the number of points of the manually generalised lines at the smaller scales (see section 4.6). If there was a large reduction in the number of points, then the tolerance would be correspondingly larger. On the other hand, the number of points is linked to the detail of the line, and therefore if a line retains much of its detail when it is manually generalised, then the automated generalisation will also be smaller.

Due to the meandering pattern of the Portuguese river, there was a large number of points, and the angles associated with this shape were very large. In addition, the meandering pattern of the river was maintained when the river was manually generalised. For this particular river, there was hardly any change of angularity between the 1 : 200 000 and the 1 : 500 000 scale maps when the lines were generalised automatically, because the number of points at the two scales were very similar (164 and 161 points respectively). As a consequence, the Douglas–Peucker tolerances needed to reach this number of points were very similar (6.5 and 6.8 m; see Appendix B). Therefore, the tolerance used was not large enough to straighten the line. In contrast to this, the Portuguese road lost much more detail when it was manually generalised. This decrease in angularity was also evident when automatically generalising the line, because the number of points was reduced to a large extent (from 52 to only 15 points). This led to a large increase in the Douglas–Peucker tolerance, from a 3.65 m wide band in the case of the 1 : 200 000 scale to 50 m for the 1 : 500 000 scale (see Appendix B). The loss of detail reflected the larger point reduction that determined the large difference in the tolerance.

The features that showed the largest difference in angularity change between the manual and the automated generalisation were the Portuguese county boundary and one of the British railway lines. The manual generalisation of the Portuguese boundary was especially drastic, and the feature lost most of its detail at the scale of 1 : 500 000. Although the initial angularity of this feature was high, the cartographer appeared to have cut straight through the feature's detail and there was a large decrease in angularity. This was reflected, for example, in the loss of length, which was the highest of all feature types (see Figure 5.7). This did not happen when the boundary was generalised automatically. In this case, most of the angles were retained, because a very large tolerance would have been necessary for the Douglas–Peucker algorithm to eliminate them.

In the British case, the railway to Ramsgate was the feature that showed the largest difference in angularity between the manual and the automated generalisation. Of the manually generalised lines for Britain, the railway was the only one to have an increase of angularity with increasing scale, as can be seen from Figure 5.18. The increase in angularity was due to bends in the railway to Ramsgate being exaggerated at the scale of 1 : 250 000. When the line was generalised automatically, the Douglas–Peucker algorithm could not add detail to the line (due to the way in which it operates) and therefore the increase was no longer observed (see Figure 5.19).

In summary, line length normally decreased but occasionally increased with scale reduction. This was dependent not only on feature type but also on particular circumstances; for example, as the circular road was enlarged, the slip road decreased. Overall, the Portuguese data were more affected than the British data, partly due to the severe generalisation of the boundary. Automatic generalisation of features retained their lengths almost entirely and caused much less distortion than manual generalisation. Sinuosity had a variation similar to that of line length with the change of scale. It usually decreased with the reduction of scale (i.e. the lines became less 'wiggly') but sometimes increased (e.g. when particular bends were exaggerated). If only portions of the features were displaced (e.g. to move around Canterbury), then this would lead to an increase in both length and angularity, because it would introduce a new shape to the line which was not there previously. Again, automatic generalisation was better at preserving angularity than manual generalisation.

5.3   Feature elimination

One of the most striking aspects of map generalisation is the elimination of features (see Figures 5.1–5.6). The number of features eliminated, and therefore the completeness of a GIS database, can be measured by retrieval operations. Elimination of features was measured in this study by making a selective search of all the rivers and all the roads for both the countries at the different source scales. In the study areas of both countries, the main features that were eliminated were the roads and the rivers. For both countries, the railways and boundaries were retained on the smaller-scale maps (with the exception of the British district boundary, which was *not shown* on the 1 : 250 000 scale map). In other words, there were no 'minor' railways or boundaries on the maps. Elimination is most often used when the features have an hierarchical organisation, such as in the case of roads and rivers.

Since the Portuguese data were obtained as paper maps and had to be digitised by the author, it was possible to keep control of the labelling of individual features (e.g. a series of disparate pieces of road on the largest-scale map were aggregated as one feature if that is how they appeared on the smallest-scale map). This approach was not practical with the British data, as they had already been digitised, and it was difficult to keep track of *all* of the individual features, many of which were very fragmented. For this reason, it was practical to count the number of features present only in the Portuguese study area. The numbers and lengths of rivers and roads for the three Portuguese scales are shown in Table 5.4. It can be seen that there was a sharp decrease in the number of features and the total length from one representation to the next, less detailed map, especially for rivers.

Table 5.4   The numbers and lengths of all of the rivers and roads in the Portuguese study area.

Scales	Rivers		Roads	
	Number	Length (m)	Number	Length (m)
1 : 50 000	185	240 756	45	72 824
1 : 200 000	19	79 759	11	46 063
1 : 500 000	2	21 398	4	32 759

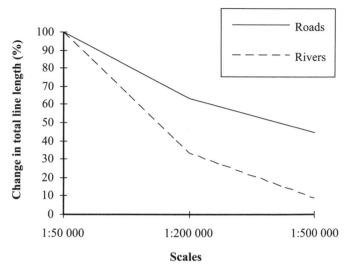

Figure 5.20 Inter-scale changes of the total line length of the rivers and roads of the Portuguese study area.

While the elimination of features was expected, it was important to determine whether the decrease followed a geometric progression. It was possible to compare the Portuguese results with the decrease in features expected from applying Töpfer's law (Töpfer and Pillewizer, 1966), which calculates the number of features suitable for depiction on a derived map, given the scale of the source map (see section 2.2.1). According to Töpfer's law, approximately 50 per cent of the features present at the scale of 1 : 50000 should be present at the scale of 1 : 200000, when in fact there were only 24 per cent of the roads and 10 per cent of the rivers left at this scale. Also, according to Töpfer's law, approximately 32 per cent of the features present on the 1 : 50000 scale map should be present on the 1 : 500000 scale map, when in fact there were only 9 per cent of the roads and 1 per cent of the rivers left.

These calculations were done *without* the inclusion of the two additional constants (the constants of symbolic exaggeration and form), because Töpfer and Pillewizer suggested that these constants were only necessary for very small scale atlas maps (i.e. scales smaller than 1 : 1 million). It is possible that the discrepancies observed were due to the Portuguese maps being of a much larger scale than the maps used by Töpfer and Pillewizer in their study. This meant that the 1 : 50000 scale map contained a large number of relatively trivial river tributaries and roads that were frequently eliminated. With smaller-scale transitions this selection has already been carried out, and so the relative decrease in feature numbers would be less extreme.

Because the eliminated roads and rivers tended to be the shorter ones, a different indicator of the elimination of features from one scale to another was used: the decrease in total feature length (see section 4.2.1). For example, between the 1 : 50 000 and the 1 : 500 000 scale Portuguese maps, 99 per cent of the *number* of rivers and 91 per cent of the *number* of roads were eliminated, but between the same scales in terms of *length*, 91 per cent of the rivers and 55 per cent of the roads were eliminated. For both feature types, the decrease in total feature length was smaller than the reduction in the total number of features. This decrease in features, as it takes account of the total length of all of the features present, is best described as a decrease in the *quantity* (rather than number) of features. The decrease in the quantity of the rivers and the roads of the Portuguese study area for the different scales is shown in Figure 5.20.

In the case of the British maps – and in contrast to the Portuguese maps – there were more roads than rivers in terms of total length (see Table 5.5). As the scale decreased, the quantity of the roads decreased more than the quantity of the rivers (see Figure 5.21). This is the opposite of what happened in the Portuguese case, and was mainly a reflection of the different geographies of both areas. Although both study areas were similar in size (the British region covered 200 km^2 and the Portuguese region 160 km^2), the British region featured 12 times the length of roads shown in the Portuguese study area at the scale of 1 : 50 000. The Portuguese region was more rural than the British one and hence there were fewer roads. Another reason for this difference might be the higher priority given to the representation of roads in relation to rivers by the OS cartographers.

The larger number of rivers in the Portuguese region (compared with the British region) was due either to the fact that there really were more rivers, or because the inferior road density allowed more space for the cartographer to draw the rivers – or a combination of both. However, as before, it might also be due partly to a higher priority given to the representation of rivers in relation to roads by the IPCC cartographers.

In summary, comparing Figures 5.20 and 5.21, it can be seen that more roads than rivers were eliminated in the British maps and that the opposite happened in the Portuguese case. This could be due to the initial amount of both feature types in both countries. As there were more roads than rivers in

Table 5.5 The lengths of all of the rivers and all of the roads of the British study area.

Scales	Length of rivers (m)	Length of roads (m)
1 : 50 000	79 296	868 661
1 : 250 000	42 086	350 970
1 : 625 000	16 180	105 835

N.B. It was not practical to count the number of features across the different scales due to the fragmentation of the features in the OS digital maps.

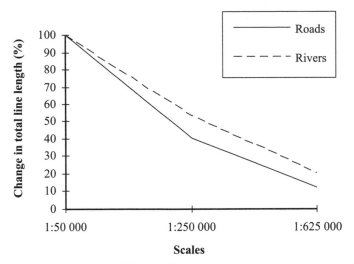

Figure 5.21   Inter-scale changes of the total line length of the rivers and roads of the British study area.

the British maps, in relative terms more roads were eliminated. Conversely, in the Portuguese maps, because there were more rivers than roads initially, more rivers were eliminated. At least for these two regions, the greater the initial quantity of features of a particular feature type, the greater the degree of elimination.

5.4   Lateral displacement of features

While the length and angularity measurements gave an insight into how the geometric characteristics of the individual features changed, and the measurements of the quantity of features gave an indication as to how the features were eliminated from one scale to another, another measurement was still necessary to give a complete picture of the effects of generalisation. This was the measurement of how individual features shifted with scale change; that is, how the positional accuracy of the lines changed. The measurement of the sideways shift of features due to generalisation complements the results shown so far, and was instrumental in the interpretation of the impact of generalisation on the GIS map manipulations.

In order to distinguish when displacement of the features was caused by generalisation rather than random errors, it was necessary to take into account the non-generalisation error of the maps. As described in section 4.1, for this study the cut-off point beyond which the error was assumed to be more likely to be due to generalisation is 0.4 mm for all of the smaller-scale maps of both countries. Below this cut-off point the error could be due either to random errors or to generalisation. However, as the size of the error increases there is an increased likelihood that this is due to generalisation, unless there are large undetected errors which were not corrected during the map-making process (an unlikely event; see section 4.1). For the different map scales the cut-off point assumed a different size in metres on the Earth's surface. The conversion of the 0.4 mm cut-off point to ground distances for the different map scales is shown in Table 5.6.

So that the sideways shift of the features caused by generalisation could be measured, the features from the smaller scale maps were enlarged and overlaid with the equivalent features represented on the largest-scale map (1 : 50 000 for both countries). The measurement of the displacement of the lines is therefore based on the assumption that the 1 : 50 000 map is less generalised than the smaller-scale maps. As explained in section 4.2.1, by comparing the features in pairs it was possible to measure the difference in position of the features and, therefore, find the sideways shift that the lines suffered when they were generalised. The next section explains how the same measurements were made for the different feature classes for both countries.

5.4.1   The shift of features on manually generalised maps

By viewing the displacement pictorially, the extent of the displacement for the particular features could be qualitatively analysed. It was then possible to use these 'displacograms' to pin-point areas of most displacement, confirmed by a quantified measurement – the maximum vector displacement. In addition to this value, the *average* displacement of a feature was determined by measuring the area between the original line and the line at the smaller scale. The number of polygons generated by the displacement and the shift of well-defined points (i.e. intersection points between features) were also measured. The total number of polygons and the total areal displacement were divided by the length of each feature (measured at the largest scale), so that the values for the

Table 5.6   Conversion of the cut-off point from mm on the map sheet to metres on the terrain.

Map scales	Size in metres of 0.4 mm on the map sheet
1 : 200 000	80 m
1 : 250 000	100 m
1 : 500 000	200 m
1 : 625 000	250 m

different feature types and for the two countries were standardised and hence comparable (see section 4.2.1).

Displacement of the Portuguese features

The displacement of the different Portuguese features between the 1 : 50 000 and the two smaller-scale maps is shown in Figures 5.22–5.25. The displacement suffered by the Portuguese railway for both the 1 : 200 000 and the 1 : 500 000 scale maps is illustrated in Figure 5.22. There was much less displacement for the 1 : 200 000 scale map, and the displacement that did exist was constant along most of its length. For the 1 : 500 000 scale map, the displacement is more pronounced and also more variable. Larger displacements at this scale occurred at point 1, where there was a loop in the railway, and also along straighter sections of the railway (e.g. at point 2). In contrast, other sections of the railway suffered little displacement (e.g. point 3).

The increased displacement of the roads with the reduction of scale is highlighted in Figure 5.23. While some portions of the roads at the scale of

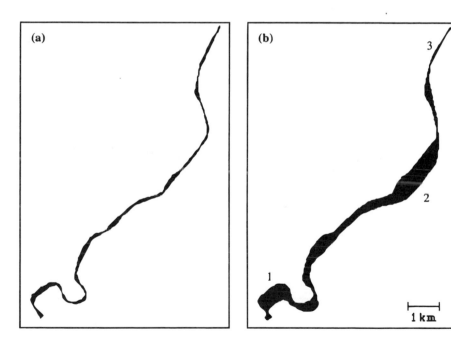

Figure 5.22 The sideways shift of the Portuguese railway. (a) The displacement of the railway at the scale of 1 : 200 000 when compared with the same railway represented at the scale of 1 : 50 000. (b) The displacement of the railway at the scale of 1 : 500 000 when compared with the same railway represented at the scale of 1 : 50 000. The numbers marked on the displacograms refer to specific instances discussed in the text.

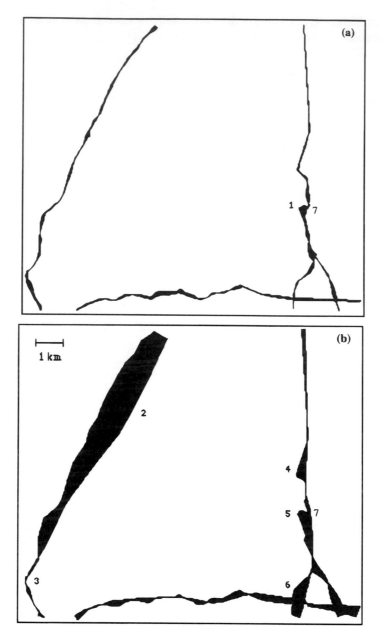

Figure 5.23   The sideways shift of the Portuguese roads. (a) The displacement of the roads at the scale of 1 : 200 000 when compared with the same roads represented at the scale of 1 : 50 000. (b) The displacement of the roads between the scales of 1 : 50 000 and 1 : 500 000. The numbers marked on the displacograms refer to specific instances discussed in the text.

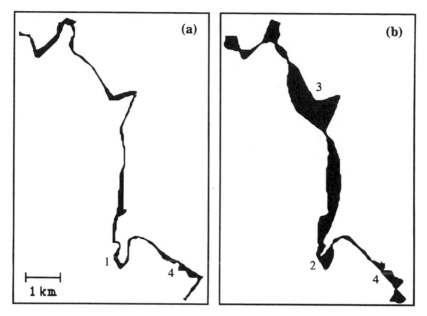

Figure 5.24   The sideways shift of the Portuguese boundary. (a) The displacement of the boundary at the scale of 1 : 200 000 when compared with the same boundary represented at the scale of 1 : 50 000. (b) The displacement of the boundary between the scales of 1 : 50 000 and 1 : 500 000. The numbers marked on the maps refer to specific instances discussed in the text.

1 : 200 000 showed some displacement (e.g. point 1, where the loop is cut off), most displacement occurred on the smaller-scale map. Point 2 in Figure 5.23 shows a large displacement of one of the roads, which contrasts with the limited displacement of the same road at point 3. The reason for this is that at point 3 there is the town of Figueira e Barros, which forces the road into a more accurate position. The location of the town, as a fixed point, therefore acts to minimise the displacement of this road. This is the opposite of what often happens when features in close proximity are displaced because of the lack of map space.

The large displacement of the Portuguese road (road with code 36) at point 2 (see Figure 5.23) could not be explained by the need to avoid features (there are no nearby features) or to simplify the detail of the line (the line is quite straight at the scale of 1 : 50 000). In this case, the displacement of the line did not appear to be straightforward generalisation and might have been partly caused by a large accidental gross error. This is unlike points 4, 5 and 6, which were displaced to simplify the detail. Point 7 in both scales shows the point of intersection of the road 245 with the main river, the Ribeira Grande. Of all the intersection points, this intersection had the smallest amount of displacement at both scales – the cartographer had probably ensured that the road accurately crossed the river where the bridge was located.

The displacement of the Portuguese boundary is shown in Figure 5.24. At point 1 on the 1 : 200 000 scale map, the boundary closely follows the line of the Ribeira Grande. However, at the same location on the smaller-scale map (point 2), the boundary now cuts across the river. Similarly, at point 3 at the scale of 1 : 500 000 the cartographer cuts off most of the detail that existed at the scale of 1 : 50 000. At point 4, on both smaller scales, the boundary follows the other river, the Ribeira de Lupe, which has become straighter, and therefore there is a corresponding loss of detail of the boundary.

The displacograms of the Ribeira Grande and the Ribeira de Lupe are shown in Figure 5.25. While there is little displacement on the 1 : 200 000 scale map, the loops of the longer river (the Ribeira Grande) on the 1 : 500 000 scale map have been simplified, while still retaining some aspects of the curves. The cartographer maintained the meandering pattern of the river while losing positional accuracy. In the case of the Ribeira de Lupe (the smaller tributary), the river lost most of its detail at the scale of 1 : 200 000 and therefore there was only some sort of *local* displacement. At the scale of 1 : 500 000, besides the

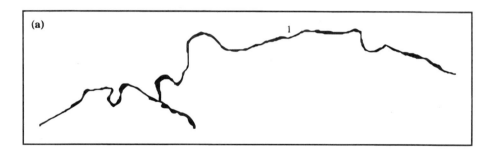

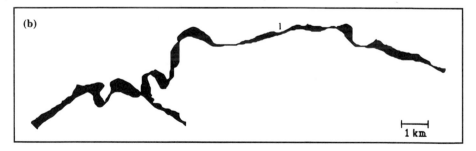

Figure 5.25 The sideways shift of the Portuguese rivers. (a) The displacement of the rivers at the scale of 1 : 200 000 when compared with the same rivers represented at the scale of 1 : 50 000. (b) The displacement of the rivers between the scales of 1 : 50 000 and 1 : 500 000. The number marked on the displacograms refers to specific instances discussed in the text.

Table 5.7 The maximum sideways shift per feature class for the Portuguese study area.

Feature class (no. of features inside brackets)	Between 1 : 50 000 and 1 : 200 000 scale maps			Between 1 : 50 000 and 1 : 500 000 scale maps		
	Max. within feature class	Mean within feature class	Std. dev.	Max. within feature class	Mean within feature class	Std. dev.
Railway (1)	131 m (0.7 mm)	n.a.	n.a.	538 m (1.1 mm)	n.a.	n.a.
Roads (4)	260 m (1.3 mm)	161 m (0.8 mm)	68 m (0.3 mm)	983 m (2.0 mm)	473 m (0.9 mm)	348 m (0.7 mm)
Rivers (2)	177 m (0.9 mm)	161 m (0.8 mm)	23 m (0.1 mm)	516 m (1.0 mm)	373 m (0.7 mm)	203 m (0.4 mm)
Boundary (1)	155 m (0.8 mm)	n.a.	n.a.	1110 m (2.2 mm)	n.a.	n.a.

Table 5.8 The displacement of intersection points between the features at the scales of 1 : 50 000 and 1 : 200 000.

Features	Road ac 1 (245)	Road ac 35	Road ac 41 (243)	Railway	River ac 32 (Rib. Grande)	River ac 175 (Rib. de Lupe)
Road ac 1 (245)						
Road as 35						
Road ac 41 (243)	79.4 m (0.40 mm)	93.1 m (0.46 mm)				
Railway		101.3 m (0.51 mm)	71.4 m (0.36 mm)			
River ac 32 (Ribeira Grande)	12.9 m (0.06 mm)			92.0 m (0.46 mm)		
River ac 175 (Ribeira de Lupe)			46.3 m (0.23 mm)		58.6 m (0.29 mm)	

simplification of the shorter Ribeira de Lupe, this river was also displaced along its entire length and therefore suffered both *local* and *global* displacement. Point 1, at both scales, shows the point of intersection of the Ribeira Grande with the road 245, which had the least displacement of all feature intersections within the Portuguese study area.

Data on the maximum sideways shift for the Portuguese features are given in Table 5.7. The maximum value within each feature class, as well as the mean maximum displacement and standard deviation, are presented (the maximum shift for each individual feature can be found in Appendix B). Comparing Table 5.7 with the values presented in Table 5.6, it can be concluded that all features have maximum and *mean* displacement values greater than the 0.4 mm cut-off point. For all of the feature classes, the maximum observed sideways shift happened on the 1 : 500 000 scale map.

From Table 5.7, it can also be seen that the feature class with the higher displacement at the scale of 1 : 200 000 was not necessarily the one with more displacement at the scale of 1 : 500 000. For example, rivers have a higher maximum displacement than the railway at the scale of 1 : 200 000, but the inverse happens at the scale of 1 : 500 000. Also, there was larger variation within classes for the smaller-scale map. With the exception of the boundary at the scale of 1 : 500 000, which at a certain point showed a very large displacement (greater than 1 km), the roads were the features that were subject to larger displacements at both scales. The high maximum displacement of the road feature class at the scale of 1 : 500 000 largely resulted from the displacement of a particular road (the road with code 36) which, as explained above, could have suffered some displacement due to a gross error.

Although the maximum vector displacement measures the maximum displacement for each feature class, it does not necessarily measure exactly the same point across the scales. Therefore, in addition to this descriptor, the measurement of the displacement of well-defined points (such as road intersections) was added. This has the advantage that the displacement of exactly the same point is measured across scales. In Tables 5.8 and 5.9 are shown the displacements of the intersection points between all of the Portuguese features (with the exception of the road with code 36, which does not intersect any other feature within the study area) for the scales 1 : 200 000 and 1 : 500 000 respectively.

It is interesting to note that for both scales the minimum displacement occurred between the main river (the Ribeira Grande) and the main road (road 245). It can be seen in the tables that the displacement of this intersection is equal to 0.06 mm in the case of the 1 : 200 000 scale map and 0.15 mm for the 1 : 500 000 scale map. These values are much inferior to the cut-off point of 0.4 mm. In the case of the 1 : 200 000 scale map, half of the intersection points were inferior to this value, and therefore only a few of the intersections were clearly affected by generalisation. In contrast, in the case of the 1 : 500 000 scale map most intersections had a displacement larger than the 0.4 mm

Table 5.9 The displacement of intersection points between the features at the scales of 1 : 50 000 and 1 : 500 000.

Features	Road ac 1 (245)	Rd ac 35	Road ac 41 (243)	Railway	River ac 32 (Rib. Grande)	River ac 175 (Rib. de Lupe)
Road ac 1 (245)						
Road ac 35						
Road ac 41 (243)	416.6 m (0.83 mm)	464.5 m (0.93 mm)				
Railway		680.9 m (1.36 mm)	229.9 m (0.46 mm)			
River ac 32 (Ribeira Grande)	74.5 m (0.15 mm)			189.6 m (0.38 mm)		
River ac 175 (Ribeira de Lupe)			715.0 m (1.43 mm)		336.8 m (0.67 mm)	

cut-off point and were therefore more severely affected by generalisation.

The *average* displacement was also calculated, to complement the maximum sideways shift and the displacement of well-defined points. The variation of the total areal displacement per unit length of each feature for the Portuguese maps is shown in Figure 5.26. The features at the smallest scale had a much higher value of total displacement. However, the feature type having the highest areal displacement value at 1 : 500 000 was not necessarily the same as that having the highest value at 1 : 200 000. This was equivalent to what was observed in the case of the maximum sideways shift of the different feature classes. It was also evident that there is greater variation between feature classes at the scale of 1 : 500 000.

At both scales, the feature type that showed the least mean displacement was the rivers. In the case of the 1 : 200 000 scale map, all features showed an average sideways shift less than the cut-off point of 80 m. The feature type that suffered the most areal displacement at the scale of 1 : 200 000 was the administrative boundaries where, for each metre of length of the boundaries at the largest scale, there was 55 m² of total areal displacement. It can be seen from Figure 5.26 that both the roads and the railway at the scale of 1 : 500 000 had an average areal displacement above the cut-off point of 0.4 mm, and hence this was probably due to generalisation. The value of areal displacement for the roads for the 1 : 500 000 scale map is 251 m² per metre of length of the road (equivalent to an average sideways shift of 251 m).

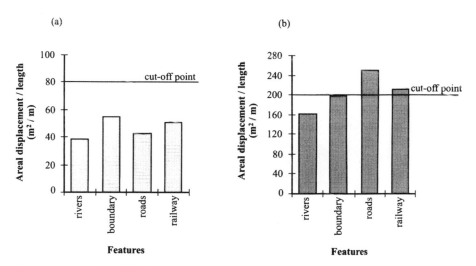

Figure 5.26 The areal displacement per unit length for each feature class of the Portuguese study area. (a) The displacement between the 1 : 200 000 and the 1 : 50 000 scale maps. (b) The displacement between the 1 : 500 000 and the 1 : 50 000 scale maps.

The roads and the railway lines showed the highest areal displacement at the scale of 1 : 500 000, although these feature classes suffered the least change in length (see Figure 5.7). The reason for this might have to do with the fact that roads and railways suffered more of a global displacement, with large sections of the features being displaced without losing much of their original shape. This can be observed particularly for the railways (see Figure 5.22), in which, for the scale of 1 : 500 000, it is the feature that generates the least number of polygons per unit length and at the same time has one of the largest average polygon sizes (see Appendix B).

The number of polygons per unit length can be a useful indicator of local versus global displacement. The more polygons per unit length a feature generates, the more it has probably lost local detail rather than suffered a global shift. The number of polygons generated by the displacement of the lines is shown in Figure 5.27. The feature type generating the largest number of polygons was the rivers at the scale of 1 : 200 000. At this scale, when overlaid with the same rivers at the scale of 1 : 50 000, the Ribeira Grande and the Ribeira de Lupe generated a total of 38 polygons, giving a mean value of 1.6 polygons per kilometre of length of the rivers at the largest scale. At the scale of 1 : 500 000, both the boundary and the roads generated the most polygons per unit length (0.7 polygons per kilometre).

The number of polygons generated was highest when comparing the 1 : 50 000 scale map with the 1 : 200 000 scale. This corroborates the previous work by Goodchild (1980a), in which he observed that lines that were less generalised (because they tended to have more detail) generated larger

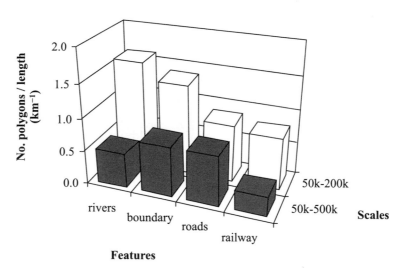

Figure 5.27 The number of polygons per unit of length of each feature for the Portuguese study area.

numbers of polygons. Although the lines from the larger-scale map generated a larger number of polygons, the polygons tended to be much smaller than the larger but fewer polygons of the small-scale map. This explains the differences between Figures 5.26 and 5.27 in which, although the 1 : 200 000 scale map generates a larger number of polygons, the total areal displacement is smaller.

Displacement of the British features

The displacement of the different British features between the 1 : 50 000 and the two smaller-scale maps is shown in Figures 5.28–5.31. Because the boundaries are not shown on the 1 : 250 000 scale map, only the displacement of the boundaries for the 1 : 625 000 scale map is shown in Figure 5.28. Although some loss of detail due to generalisation is apparent, the large displacement observed is probably not just a case of line simplification. The 1 : 625 000 map was an amalgamation of the Routeplanner, administrative and topographic OS maps. It is possible that there was a mismatch between the administrative map and the Routeplanner that partly caused this displacement.

The probable mismatch between the administrative and the Routeplanner maps can be explained by considering the life history of these data, and the fact that when digitised they were on a different base (i.e. one on plastic and the other on paper). Both the Routeplanner and the administrative map were derived from the same source map (the ten-mile map of Great Britain) but the

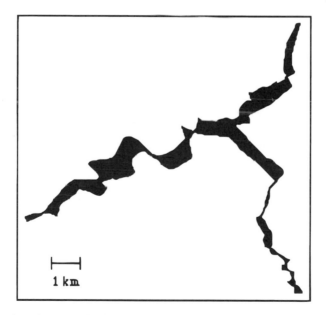

Figure 5.28   The sideways shift of the British boundaries at the scale of 1 : 625 000 when compared with the same district boundaries represented at the scale of 1 : 50 000.

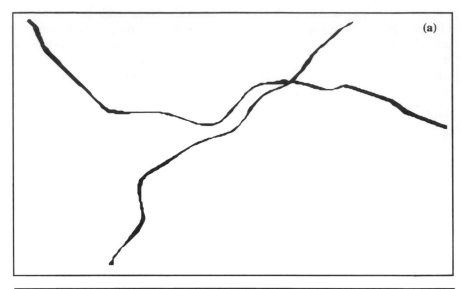

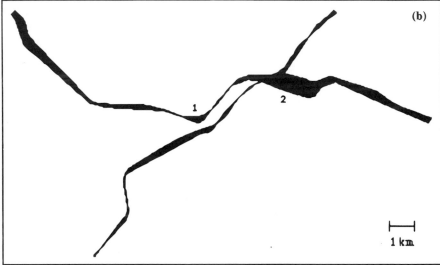

Figure 5.29 The sideways shift of the British railways. (a) The displacement of the railways at the scale of 1 : 250 000 when compared with the same railways represented at the scale of 1 : 50 000. (b) The displacement of the railways between the scales of 1 : 50 000 and 1 : 625 000. The numbers marked on the displacograms refer to specific instances discussed in the text.

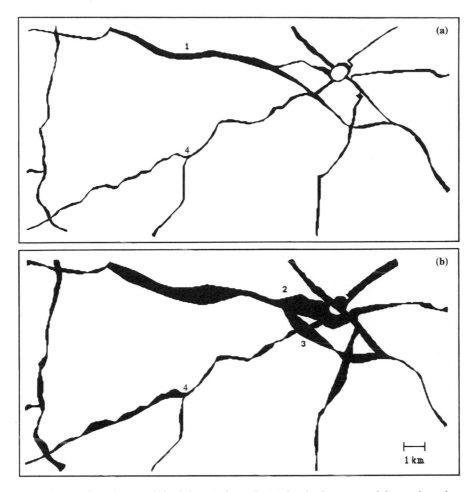

Figure 5.30 The sideways shift of the British roads. (a) The displacement of the roads at the scale of 1 : 250 000 when compared with the same roads represented at the scale of 1 : 50 000. (b) The displacement of the roads between the scales of 1 : 50 000 and 1 : 625 000. The numbers marked on the displacograms refer to specific instances discussed in the text.

maps then diverged independently. In 1978, only the Routeplanner was re-drawn and re-adjusted when the plastic version was prepared (because it had been stored on paper until that point). They were both digitised separately and later, when they were added together, an attempt was made to match the administrative map with the Routeplanner (Neil Smith, OS, 1993, personal communication). This partly explains the observed displacement.

The displacograms of the two railway lines on the British maps are shown in Figure 5.29. The two British railway lines showed little overall displacement

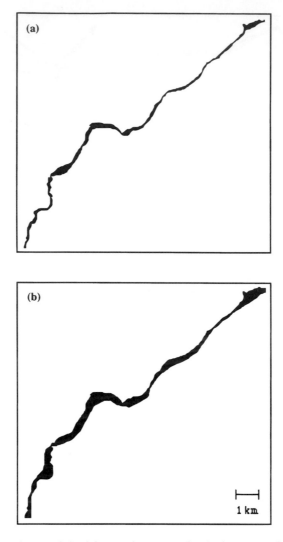

Figure 5.31   The sideways shift of the British river. (a) The displacement of the river at the scale of 1 : 250 000 when compared with the same river represented at the scale of 1 : 50 000. (b) The displacement of the river between the scales of 1 : 50 000 and 1 : 625 000.

at the scale of 1 : 250 000, but slightly more displacement was apparent at the smallest scale (see Figure 5.29). Point 1 showed a typical exaggeration of a bend, while point 2 showed where the railway was displaced southwards to curve around the enlarged circular road of the city of Canterbury. The large displacement of the section of the railway at point 2 led to the displacement of the Canterbury East railway station by 703 m.

Table 5.10 The maximum sideways shift per feature class for the British maps.

Feature class (no. of features inside brackets)	Between 1 : 50 000 and 1 : 250 000 scale maps			Between 1 : 50 000 and 1 : 625 000 scale maps		
	Max. within feature class	Mean within feature class	Std. dev.	Max. within feature class	Mean within feature class	Std. dev.
Railways (2)	144 m (0.6 mm)	135 m (0.5 mm)	13 m (0.1 mm)	701 m (1.1 mm)	469 m (0.8 mm)	328 m (0.5 mm)
Roads (13)	323 m (1.3 mm)	139 m (0.6 mm)	65 m (0.3 mm)	927 m (1.5 mm)	410 m (0.7 mm)	245 m (0.4 mm)
River (1)	256 m (1.0 mm)	n.a.	n.a.	488 m (0.8 mm)	n.a.	n.a.
Boundaries (3)	n.a.	n.a.	n.a.	910 m (1.5 mm)	662 m (1.1 mm)	248 m (0.4 mm)

Table 5.11 The displacement of intersection points between features at the scales of 1 : 50 000 and 1 : 250 000.

Features	Rd ac 8 (A251)	Rd ac 9 (A2)	Rd ac 11 (M2)	Rd ac 12 (A252)	Rd ac 13 (A28)	Rail ac 2 (to Ramsgate)	Rail ac 3 (to Dover)	Rv ac 24 (Great Stour)
Rd ac 8 (A251)								
Rd ac 9 (A2)								
Rd ac 11 (M2)	82.4 m (0.33 mm)							
Rd ac 12 (A252)	104.3 m (0.42 mm)							
Rd ac 13 (A28)		196.8 m (0.79 mm)		47.5 m (0.19 mm)				
Rail ac 2 (to Ramsgate)		120.7 m (0.48 mm)			127.2 m (0.51 mm)			
Rail ac 3 (to Dover)		230.7 m (0.92 mm)	139.7 m (0.56 mm)		152.9 m (0.61 mm)	135.9 m (0.54 mm)		
Rv ac 24 (Great Stour)		142.2 m (0.57 mm)			95.9 m^a (0.38 mm)	23.0 m^b (0.09 mm)	96.4 m (0.39 mm)	

[a] In the study area, the Great Stour intersects the A28 in three places – this value is the one with the minimum displacement.
[b] In the study area, the river intersects the railway to Ramsgate in three places – this value is the one with the minimum displacement.

The displacement of the roads is shown in Figure 5.30. For the 1 : 250 000 scale map there is one section of a road which has a displacement that stands out: this is the A2 trunk road at point 1 (refer to the displacogram in Figure 5.30a). This sideways shift might be due to the fact that this road was built after the paper map had been compiled, and was therefore probably digitised at a different time. When updating new features on the smaller-scale maps, the cartographer can be particularly concerned to maintain the relative accuracy of the features rather than the absolute accuracy in relation to the National Grid. This is because topological accuracy is often more important on small-scale maps (Neil Smith, OS, 1993, personal communication).

Point 2 on the 1 : 625 000 scale map (see Figure 5.30b) is the slip road that had been straightened and shortened. Point 3 is the bypass around Canterbury, which was shifted to avoid overlapping the railway (which in turn had been shifted southwards by the enlargement of the road that encircles the city of Canterbury – see above). This shows how the generalisation of one feature can have a knock-on effect on several others. At both the scales of 1 : 250 000 and 1 : 625 000, the minimum displacement of a road intersection was observed at point 4. This is the intersection between the A28 and the A252.

Finally, the displacement of the river for the British maps is shown in Figure 5.31. In the case of the scale of 1 : 250 000, the overall displacement seems to have been fairly minimal. The smaller-scale map had the most sideways shift, due mainly to the simplification of the line and the cutting of corners. The cutting of corners caused quite large local displacements at the scale of 1 : 625 000, with the creation of a small number of large polygons.

In Table 5.10 are shown the data related to the sideways shift for the British features: the maximum value within each feature class, the mean maximum displacement and the standard deviation. Comparing Table 5.10 with the values presented in Table 5.6, it can be concluded that all of the features had maximum displacement values greater than the 0.4 mm cut-off point. It can also be seen that the feature class with a higher displacement at the scale of 1 : 250 000 was not necessarily the one with the most displacement at the scale of 1 : 625 000. As in the Portuguese case, the roads were the features that were subject to larger displacements at both scales, with the exception of the boundaries. The variation within classes was greatest for the smaller-scale map. All of these were very similar to the results for the Portuguese data. Unexpectedly, however, the river actually had *less* maximum displacement (measured in millimetres on the map) in the case of the scale of 1 : 625 000 than was present at the scale of 1 : 250 000.

While the maximum vector displacement pin-pointed the sections of the lines that showed the largest displacement, in Tables 5.11 and 5.12 are shown the displacements of intersection points for the scales 1 : 250 000 and 1 : 625 000. The following roads were used for this analysis: the M2 motorway, the dual carriageway, the A28 and the A251. As in the case of the Portuguese study area, the intersection with the minimum displacement was the same for

Table 5.12 The displacement of intersection points between features at the scales of 1 : 50 000 and 1 : 625 000.

Features	Rd ac 8 (A251)	Rd ac 9 (A2)	Rd ac 11 (M2)	Rd ac 12 (A252)	Rd ac 13 (A28)	Rail ac 2 (to Ramsgate)	Rail ac 3 (to Dover)	Rv ac 24 (Great Stour)
Rd ac 8 (A251)								
Rd ac 9 (A2)								
Rd ac 11 (M2)	376.6 m (0.60 mm)							
Rd ac 12 (A252)	259.8 m (0.42 mm)							
Rd ac 13 (A28)		799.8 m (1.28 mm)		256.7 m (0.41 mm)				
Rail ac 2 (to Ramsgate)		557.0 m (0.89 mm)			283.4 m (0.45 mm)			
Rail ac 3 (to Dover)		412.8 m (0.66 mm)	333.7 m (0.53 mm)		994.2 m (1.59 mm)	708.3 m (1.13 mm)		
Rv ac 24 (Great Stour)		645.2 m (1.03 mm)			432.3 m[a] (0.69 mm)	219.8 m[a] (0.35 mm)	658.8 m (1.05 mm)	

[a] The same intersection point as in the case of the scale of 1 : 250 000 was used.

both scales and it also involved a river. This happened at one of the points at which the Great Stour crossed the railway to Ramsgate. In the case of the scale of 1 : 250 000, the displacement of this intersection was 0.09 mm, while for the scale of 1 : 625 000 the displacement was 0.35 mm. In both cases, the displacement was inferior to the cut-off point of 0.4 mm and it was therefore not possible to attribute the displacement to generalisation or other map error. For the 1 : 625 000 scale map this was the *only* intersection point that had a value inferior to 0.4 mm. In the case of the scale of 1 : 250 000, four other intersection points had values inferior to the 0.4 mm cut-off point. Two of these also involved intersections with the Great Stour, and the other two were road intersections.

In Figure 5.32 is shown, for the British maps, the areal displacement (per metre length) of each feature. As in the Portuguese case, the highest total areal displacement was produced at the scale of 1 : 625 000 (in relation to the scale of 1 : 50 000). From the figure, it can be seen that the boundaries generated the largest displacement. There was 287 m² areal displacement for each metre of length of the boundaries at the scale of 1 : 625 000 when compared with the 1 : 50 000 scale map (equivalent to an average sideways shift of 287 m). Only the boundaries at the scale of 1 : 625 000 had an average sideways shift greater than the 0.4 mm cut-off point (equivalent to 250 m for the 1 : 625 000 scale map). However, as explained earlier, this was probably also due to error in matching the different source maps when producing the digital data, rather than being exclusively due to generalisation.

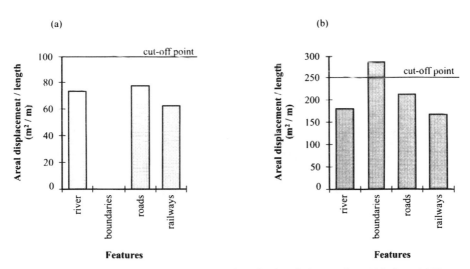

Figure 5.32 The areal displacement per unit length of each feature for British data. (a) The displacement between the 1 : 250 000 and the 1 : 50 000 scale maps – there are no data for the boundaries. (b) The displacement between the 1 : 625 000 and the 1 : 50 000 scale maps.

It could not be determined whether the boundaries were also the features that showed the largest sideways shift at the scale of 1 : 250 000, as at this scale the boundaries were not represented. At 1 : 250 000, the roads were the feature class that had the largest areal displacement. The roads had 78 m² displacement per metre of road at the largest scale (equivalent to an average sideways displacement of 78 m). None of the feature classes at the scale of 1 : 250 000 had an average sideways shift larger than the 0.4 mm cut-off point (i.e. none were larger than 100 m). The features that had the least mean displacement at both scales were the railways.

The number of polygons generated per unit length of the feature at the largest scale is shown in Figure 5.33. In this case, and in contrast to Figure 5.32, the largest values were for the 1 : 250 000 scale map when comparing this map with the 1 : 50 000 scale, because lines with more detail generate larger numbers of polygons. As for the Portuguese features, because of the higher angularity, rivers were the feature type that generated the largest number of polygons for the scale of 1 : 250 000. For this scale, the river generated 12 polygons, corresponding to 0.7 polygons per kilometre of length of the river at the largest scale. The reason why the boundaries generated such a high number of polygons at the scale of 1 : 625 000 was probably that at this scale the boundaries are the feature with the highest angularity (see Table 5.3). In contrast, the river generated the least number of polygons at the scale of 1 : 625 000, despite its large angularity. This is due to the way in which the cartographer cut through the river's meanders, generating a small number of large polygons (see Figure 5.31b).

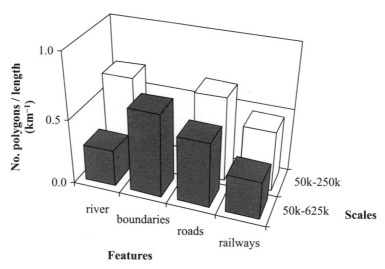

Figure 5.33 The number of polygons per unit length of each feature for the British data. Note that the district boundaries are not represented at the scale of 1 : 250 000.

5.4.2 Displacement for automatically generalised features

In order to compare the areal displacement between lines generalised manually and lines generated by automated algorithms, four features for both countries were used – one river, one road, one boundary and one railway line (the same features as used previously in sections 5.1.2 and 5.2.2). The areal displacement per unit length of the features at the largest scale for the Portuguese study area is shown in Figure 5.34. It can be seen that the areal displacement was much larger for the lines that were generalised manually than for those generalised automatically.

In Figure 5.35 it is shown that the number of polygons generated from lines generalised automatically was much larger than when manual methods were used to generalise the lines. This was to be expected, as the lines that are generalised automatically have only a slight reduction in detail and therefore will tend to generate a large number of polygons with a small area. Automatic generalisation caused more frequent but less significant differences. In the case of the lines generalised automatically, the polygons were *sliver* polygons.

The areal displacement per unit length of the features at the largest scale for the British study area is shown in Figure 5.36. As in the case of the Portuguese study area, the areal displacement was much larger for the manually generalised lines than for those generalised automatically. The number of polygons generated by the British lines is shown in Figure 5.37. As with the results

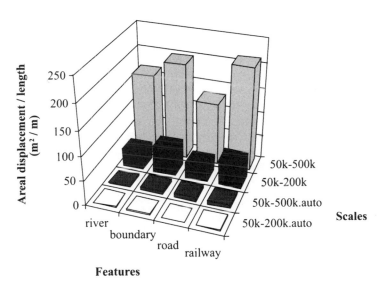

Figure 5.34 The areal displacement per unit length of each feature, generalised manually and automatically, for the Portuguese data. Note that the pairs of maps with smaller values are placed in the front rows for clarity.

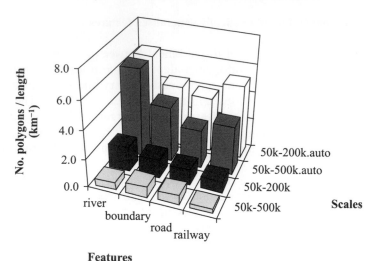

Figure 5.35   The number of polygons per unit length of each feature, generalised manually and automatically, for the Portuguese features. Note that the pairs of maps with smaller values are placed in the front rows for clarity.

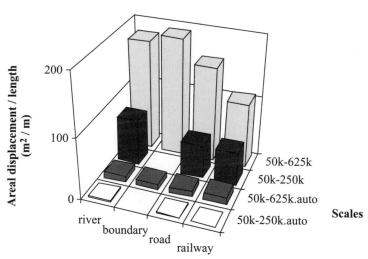

Figure 5.36   The areal displacement per unit length of each feature, generalised manually and automatically, for the British features. Note that the district boundaries are not represented at the scale of 1 : 250 000, and that the pairs of maps with smaller values are placed in the front rows for clarity.

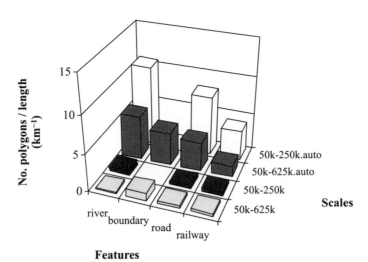

Figure 5.37   The number of polygons per unit length of each feature, generalised manually and automatically, for the British data. Note that the district boundaries are not represented at the scale of 1 : 250 000, and that the pairs of maps with smaller values are placed in the front rows for clarity.

obtained for the Portuguese study area, the number of polygons generated from lines generalised automatically was much larger than when manual methods were used to generalise the lines.

In summary, while the displacement of features was variable (dependent on feature type, geographical position, length, and proximity to other generalised features) there were a few general trends. As expected, the displacement usually became greater as the scale of the map became smaller. The Portuguese features that were consistently displaced the least were the rivers. This corroborates the efforts of their cartographers, who try to retain the positional accuracy of rivers in preference to other features (see section 3.1.2). In contrast, the railways in the British maps were displaced the least at both scales. While large displacements were probably due to generalisation, there were a couple of cases (one of the Portuguese roads and the British boundaries) in which a displacement larger than the 0.4 mm cut-off point could have been partly caused by gross errors rather than generalisation. The automatic generalisation methods were more effective than manual generalisation in minimising displacement, although they created more sliver polygons. In chapter 6, generalisation effects are analysed in the light of GIS map manipulations.

CHAPTER SIX

The impact of generalisation on GIS map manipulations

A key aspect of the research covered in this book was to investigate the effects of generalisation on GIS map manipulations, using digital data of different source scales. A point-in-polygon and an overlay operation were carried out in order to determine how generalisation affected typical GIS map manipulations. These map manipulations were simpler than most full-blown GIS analyses, and so made it easier to disentangle the effects of generalisation. The two GIS operations were typical of environmental-type applications. The point-in-polygon was for the modelling of noise impact and the overlay operation was for the siting of a new park.

For the two operations, a different set of features was used. The point-in-polygon procedure used only road data, in order to investigate a single feature type. For the overlay operation, the maximum possible number of feature types was used. This included the main line features (roads, rivers and railway lines), plus the positions of the railway stations and the outline of the urban areas. This chapter starts by describing the effects of generalisation on the point-in-polygon analysis, and this is followed by the description of the overlay operation that was carried out. The account of the overlay operation is divided into two parts. The first part covers the results of the overlay when only the features *common* to all map scales were used. The second part describes the results when the overlay was repeated using *all* of the features present on the different maps.

6.1 Effects on point-in-polygon operations

For the point-in-polygon operation carried out in this study, the region between 100 and 300 m from the roads was the search area. This sort of buffer would be appropriate for a study into noise pollution from road traffic. For example, city councils sometimes pay compensation to people living between 100 and 300 m from a new road. It is assumed by councils that houses within 100 m of a road can be protected from noise pollution by creating sound barriers, while houses beyond 300 m are not affected (Halcrow Fox, Environmental Planning Consultants, 1992, personal communication). The road buffers were derived from the roads of both countries, at the three different scales. Only the roads that were common across all the map scales were used.

In order to determine how generalisation of the search areas (rather than the change in location of the targets to be found) would affect a point-in-polygon analysis, the same set of randomly located points was used across the three map scales (see section 4.5). In addition, in order to determine whether the density of the points affected the results, two different point densities were used: *low density*, with one point per square kilometre and *high density*, with four points per square kilometre. Because the two study areas had slightly different sizes, the total number of points used was different for each country (see Table 6.1).

The points used in such a point-in-polygon analysis represented hypothetical houses in the area. Any house that fell within the buffer could be considered to be vulnerable to noise pollution. However, it should be emphasised that the random points used were theoretical and did not correspond to any real attribute from the maps used. The advantage of using artificially created points was that the difference in the results of the point-in-polygon operation was dependent exclusively on the generalisation of the road buffers, and was not due to a combination of the generalisation of the roads and the accuracy of the location of real points. The way in which the generalisation of search areas influenced the results of a point-in-polygon operation could therefore be isolated. In order to illustrate how the buffers and random points appeared, in Figures 6.1 and 6.2 are shown the low- and high-density plots for only the Portuguese 1 : 50 000 scale map. The same information for the British 1 : 50 000 scale map is shown in Figures 6.3 and 6.4.

Over the three scales, the area and position of the road buffers altered, while the position of the points remained constant. The changes in the area of the buffers were partly due to the changes in road length. As the scale

Table 6.1 The number of points used in the point-in-polygon operation for both countries.

	Total number of points used	
Point density	Portuguese study area (160 km²)	British study area (200 km²)
Low density (1 point/km²)	160	200
High density (4 points/km²)	640	800

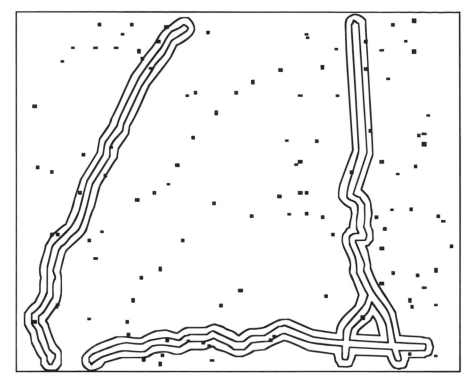

Figure 6.1 Low-density points, randomly dispersed, overlaid on the road buffers for the 1 : 50 000 Portuguese scale map.

decreased, and in relation to the 1 : 50 000 scale, the Portuguese buffers decreased slightly in size, while for the British maps the buffers increased in size (see Tables 6.2 and 6.3). This, coupled with the displacement of the roads (see section 5.4.1), influenced which points fell within the buffers. In Tables 6.2 and 6.3 are shown, for Portugal and Britain respectively, the number of points that fell within the noise-level critical region. The tables show the variation of the results with scale change and with the change of point density. As expected, when a high density of points was applied, far more points fell within the road buffers than with the low density. Also, for the British maps there was an extra 51 km of roads, and therefore the buffer area was much larger; with a corresponding increase, over the Portuguese maps, in the number of points that fell within the buffers.

It can be seen that, for both countries, the number of points that fell in the region between 100 and 300 m from the roads varied with the change in scale. This was due to the generalisation of the roads, which affected the position, size and shape of the road buffers. However, the direction and the magnitude

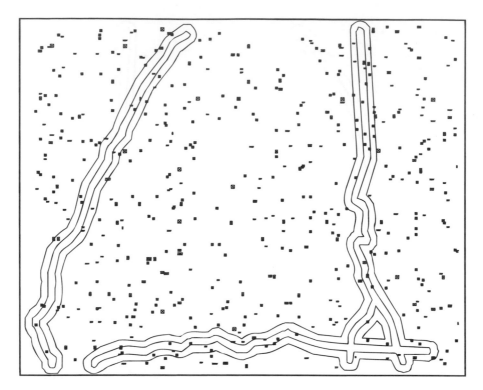

Figure 6.2   High-density points, randomly dispersed, overlaid on the road buffers for the 1 : 50 000 Portuguese scale map.

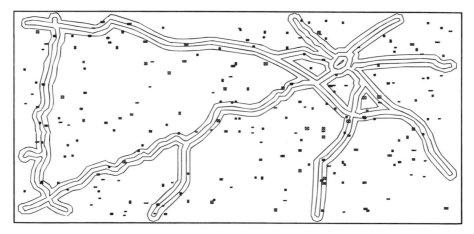

Figure 6.3   Low-density points, randomly dispersed, overlaid on the road buffers for the 1 : 50 000 British scale map.

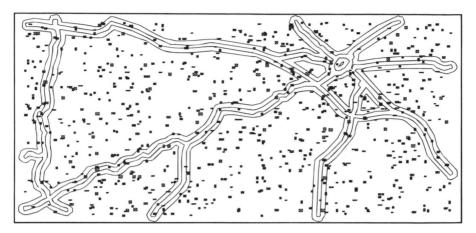

Figure 6.4 High-density points, randomly dispersed, overlaid on the road buffers for the 1 : 50 000 British scale map.

of change of the number of points with the reduction of scale for the different point densities was unpredictable. For example, in the Portuguese case, for the two point densities, the number of points decreased with the reduction of scale. For the low-density analysis, only half of the points fell within the road buffers at the smallest-scale, compared to the largest-scale map. With the British maps, while the effects tended to be smaller, the effect was more variable. For the high-density analysis the number of points consistently increased with decreasing scale, while for the low-density analysis the number of points either decreased or remained constant relative to the largest scale.

The above analysis was concerned only with the *quantitative* aspect of 'how many points'. While this is adequate if all of the points are equivalent, it is often the case that points vary in value or attribute. Therefore it is sometimes necessary to consider the *qualitative* aspects of 'which points'. All of the points had an individual code attributed to them, and it was therefore possible to determine exactly *which* points fell within 100 and 300 m of the roads. In all cases there were points that fell inside the region for one scale but not for another. In fact, the number of points that fell within the buffer regions of the three scales but which shared the same identity was considerably less than the values presented in Tables 6.2 and 6.3. In Table 6.4 is shown the number of points common to the *three* scales (i.e. with the same identity) that fell within the road buffers, for the two countries and the two point densities. It can be seen, for example, that although on average 121 points fell within the buffers of the British maps for the high density of points, only 38 points actually had the same identity. Only these were common to all three scales.

The data from both countries gave good examples of this change in the specific point membership for the buffers generated at different scales. In some

Table 6.2 The total number of points that fell between 100 and 300 m from the Portuguese roads, for the three scales and the two point densities. The area of the road buffers is in brackets.

Point density	Number of points for each scale			Variation with scale change (%)	
	1 : 50 000 (12.899 km²)	1 : 200 000 (12.745 km²)	1 : 500 000 (12.514 km²)	1 : 50 000 versus 1 : 200 000	1 : 50 000 versus 1 : 500 000
Low density (1 point/km²)	14	14	7	0	−50
High density (4 points/km²)	47	45	41	−4	−13

Table 6.3 The total number of points that fell between 100 and 300 m from the British roads, for the three scales and the two point densities. The area of the road buffers is in brackets.

Point density	Number of points for each scale			Variation with scale change (%)	
	1 : 50 000 (21.127 km²)	1 : 250 000 (21.248 km²)	1 : 625 000 (21.171 km²)	1 : 50 000 versus 1 : 250 000	1 : 50 000 versus 1 : 625 000
Low density (1 point/km²)	34	32	34	−6	0
High density (4 points/km²)	116	120	127	+3	+9

Table 6.4 Points common to the three scales (i.e with the same identity) that fell between 100 and 300 m from the roads, for the two countries and the two point densities.

| | Number of points common to all three scales | |
| | Low density (1 point/km^2) | High density (4 points/km^2) |
Country		
Portugal	5	19
Great Britain	11	38

cases, the same numbers of points fell within the areas but, as shown below, the specific points were different. For the Portuguese data, and in the case of the low density of points, the same number of points fell within the region (14 in total), for the scales of 1 : 50 000 and 1 : 200 000 (see Table 6.2). However, as can be seen in Table 6.5, of these 14 points, 12 were shared by the two scales, while the two other points were unique to each scale. This difference was accentuated even further by the 1 : 500 000 scale map, on which only six points were shared. Similar results were obtained for the high-density analysis. In the British case, and for the scales of 1 : 50 000 and 1 : 625 000, the same number of points (34 in total) fell within the road buffers (see Table 6.3). However, of these 34 points only 16 shared the same identity between the two scales (see Table 6.6). Similarly, the effect of generalisation was strong for the high-density analysis. The generalisation effect was largest for the smallest-scale maps, and this trend was consistent.

A chi square was considered an appropriate statistic to test the significance of this result, given that the distribution of the population of points was unknown. The results of the four chi square tests performed for the smaller-scale maps are shown in Table 6.7. Only the points that share the same identity as those of the 1 : 50 000 scale maps were used. The points that fell within the road buffers of the 1 : 50 000 scale map of each country were the 'expected frequency'. The difference between the different scale maps was significant (d.f. = 1, $P < 0.05$) for all four permutations. Given the uncertainty inherent in this kind of spatial distribution, this result can only be considered indicative.

It is important to emphasise that although this study revealed that there were differences in the results in a typical point-in-polygon operation due to the generalisation of the search areas, these differences could sometimes be disguised due to the points falling in and out of the increasingly generalised polygons. This in part explains the contrast in the results, for example, between Tables 6.3 and 6.6, where there were differences as little as zero per cent in one case and as large as 53 per cent in another equivalent case. The full impact of generalisation on a point-in-polygon operation is only felt when the point values within the search area are added or averaged, rather than by just totalling the number of points. In the former case, the qualitative and

Table 6.5 Points that fell between 100 and 300 m from the Portuguese roads which share the same identity as those on the 1 : 50 000 scale map.

Point density	Number of points for each scale			Variation with scale change (%)	
	1 : 50 000	1 : 200 000	1 : 500 000	1 : 50 000 versus 1 : 200 000	1 : 50 000 versus 1 : 500 000
Low density (1 point/km²)	14	12	6	−14	−57
High density (4 points/km²)	47	38	22	−19	−53

Table 6.6 Points that fell between 100 and 300 m from the British roads which share the same identity as those on the 1 : 50 000 scale map.

Point density	Number of points for each scale			Variation with scale change (%)	
	1 : 50 000	1 : 250 000	1 : 625 000	1 : 50 000 versus 1 : 250 000	1 : 50 000 versus 1 : 625 000
Low density (1 point/km²)	34	24	16	−29	−53
High density (4 points/km²)	116	77	53	−34	−54

Table 6.7   The results of four chi square tests for
the Portuguese and British point-in-polygon operations for the different-scale maps.

χ^2	Portuguese	British
Low point density	4.9	12.5
High point density	15.0	47.3

quantitative changes of the points within the search areas are combined. The
effect of generalisation on point-in-polygon analysis is a *combination* of both
the change in the total number and the specific points that lie within the
buffers.

6.2   Effects on overlay operations

For this study, an overlay operation was carried out to find an hypothetical
site for a new countryside park within each study area. The overlay analysis
used a wider range of feature types than the point-in-polygon operation. In
general terms, the required characteristics of the new park were that it be close
to a river, that it did not have a major road or railway passing through it, that
it was not within an existing urban area, and that it should have a certain
minimum size. At the same time, it was considered important that the park
should be relatively accessible by road and by railway. These typical constraints for the creation of the park were specifically as follows:

1   It should be within 500 m of a river.

2   It must not be within 200 m of a railway.

3   It must not be within 200 m of a road.

4   It should be within 5000 m of a railway station.

5   It should be within 1000 m of a road.

6   It must not be within the outline of urban areas.

7   It should be at least 500 000 m^2 in area and should be a single unit: dis-
jointed parcels summing to this area were not acceptable.

The outline of the British urban areas was shown in the digital files
obtained from the Ordnance Survey (1 : 50 000, 1 : 250 000 and 1 : 500 000
digital maps). The city of Canterbury was the only urban area boundary that
was present across all three scales. However, in the case of the Portuguese
maps, the outline of the town of Fronteira was not drawn on any of the maps.
Instead, houses (individual or aggregated in blocks) were shown at the scales
of 1 : 50 000 and 1 : 200 000, and at the scale of 1 : 500 000 the town of Fronteira became symbolised by a circle (as described in section 2.2). In order to
obtain the boundary of Fronteira, a buffer of 50 m was created around the

houses located within the town, in order to force them to coalesce (the size of the buffer being determined by the approximate distance, as drawn on the map, between the houses in the town). In this way, an approximate urban outline was formed for the scales of 1 : 50 000 and 1 : 200 000. The area occupied by the urban symbol in the 1 : 500 000 scale map was used as a surrogate for the area occupied by the town of Fronteira at this scale. As it happened, this approximation did not affect the results, because the town of Fronteira was located more than 500 m from any of the rivers.

The overlay operation was repeated three times for each country, each time using features from a different original map scale. The overlay was first done using only the features common to all map scales; in other words, features that were only present in one or two of the scales were not used. This meant that any differences across scales were due to generalisation of the same features (displacement, change in length, etc.) and not elimination of features. The overlay was then repeated using all features present in all of the maps, to simulate the result that a GIS user might typically have obtained.

6.2.1   Overlay operation using features common to all map scales

In the Portuguese case, and for the three scales, four zones were generated as a result of the overlay operation. All zones had an area greater than the minimum 500 000 m² and therefore all could *potentially* be converted into a park. Of course, the decision to select any of these areas for conversion into a park in a real case would depend on a host of extra information, such as land cover and land use, and site visits. The areas of the four zones that could be converted into a park close to Fronteira are shown in Table 6.8. The results of the overlay operation for the three Portuguese maps are shown in Figures 6.5–6.7.

From Figures 6.5–6.7 and Table 6.8, it can be seen that in each case the shape and the area of each zone changed. For nearly all of the zones, there was a decrease in the area of each zone with the reduction of scale. The exception

Table 6.8   The areas of the different potential zones for conversion into a park close to Fronteira in Portugal.

	Areas of the different zones per scale (m²)		
	1 : 50 000	1 : 200 000	1 : 500 000
Zone 1	942 748	859 228	817 658
Zone 2	944 801	864 497	810 946
Zone 3	736 868	622 939	514 629
Zone 4	829 256	708 443	918 801
Total	3 453 673	3 055 107	3 062 034

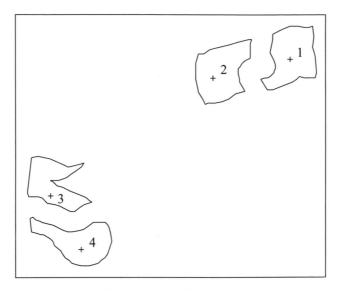

Figure 6.5 Potential zones suitable for a new park, using the 1 : 50 000 scale Portuguese map.

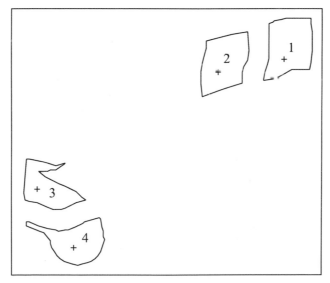

Figure 6.6 Potential zones suitable for a new park, using the 1 : 200 000 scale Portuguese map.

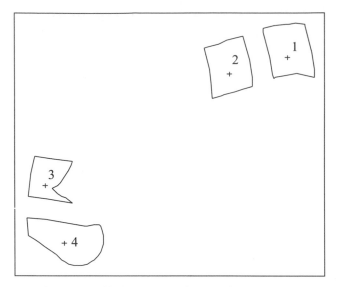

Figure 6.7　Potential zones suitable for a new park, using the 1 : 500 000 scale Portuguese map.

to this was the zone labelled 4 on the maps, which for the scale of 1 : 500 000 had an increase in area (an 11 per cent increase in relation to the scale of 1 : 50 000, equivalent to 89 545 m²). This area increase was mainly caused by the generalisation of one of the rivers and one of the roads, and in particular because of changes in their relative positions (at the 1 : 500 000 scale map, the road and the river stop running parallel to each other). In terms of the total area, there was a 12 per cent reduction of the potential area from the 1 : 50 000 to the 1 : 200 000 scale map. The decrease in area for the 1 : 500 000 scale map was smaller (only 11 per cent) due to the increase in area of zone 4.

Within the Portuguese study area, the relevant features used in the overlay (i.e. common to all three scales) were the following: two rivers, one railway, one railway station, the outline of the town of Fronteira and four roads. In the case of the Portuguese study area there were two features that did not affect the results of the overlay analysis. The town of Fronteira was located further than 500 m from the rivers and, therefore, was not within the river buffer. Also, there was no section of the railway that was located within 500 m of a river as well as being within 1000 m of a road.

For the Portuguese data, only the generalisation (in terms of displacement and change in length) of the rivers and the roads and the change in the position of the railway station contributed to the change of the shape and the area of the different zones. From section 5.4.1, it can be seen that the rivers were displaced the least, while the roads suffered more severe displacement: therefore, the latter had the greatest influence on the change in the results over scale. Despite the fact that the roads stayed more constant across scales in terms of length (while the rivers lost more length), displacement affected the relative position of the features and therefore had a larger impact on the results.

From Figures 6.5–6.7, it can be seen that the gap between zones 1 and 2 shrank and straightened out as the loop of one of the roads disappeared (equivalent to *local* displacement). The other gap between zones 3 and 4 was also caused by a road. This road straightened out and stopped following the river at the scale of 1 : 500 000. The crossing of the river by the road (perpendicular to one another) was in part the cause of the increase in size of zone 4. In the Portuguese case, the loss of detail of the roads and rivers, and the change in relative position between them, contributed to most of the variation in the size of the parks.

The position of the railway station also affected the Portuguese results. In Figure 6.8 is shown the displacement in direction and magnitude of Fronteira railway station between the 1 : 50 000 scale map and the two smaller-scale maps. On the maps at all three scales, the railway station was represented by a rectangular symbol. The precise locations of the railway station were taken as the centre of each of these rectangles. The displacement of the Portuguese railway station between the 1 : 50 000 and 1 : 200 000 scale maps corresponded to 0.5 mm in map units, and the displacement between the 1 : 50 000 and 1 : 500 000 scale maps corresponded to 0.6 mm on the map. In terms of the overlay analysis, these displacements meant that the centre of a circle with a 5 km radius was displaced, in each direction, by 92 m and 317 m for the 1 : 200 000 and the 1 : 500 000 scale maps respectively. Therefore the features, or portions of features, that were encompassed by this circle had positions that differed by as much as 317 m. This also contributed to the enlargement of area 4 in Figure 6.7.

In the case of the British study area, in Figures 6.9–6.11 are shown the maps that resulted from the overlay operation using the three Ordnance Survey maps. For the three scales, only two zones had an area larger than the minimum 500 000 m^2 (the ones numbered in the figures) and therefore could

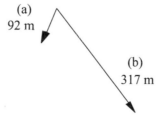

(a)
92 m

(b)
317 m

Figure 6.8   The displacement of the railway station of Fronteira. (a) The displacement between the 1 : 50 000 and 1 : 200 000 scale maps. (b) The displacement between the 1 : 50 000 and 1 : 500 000 scale maps. The vectors indicate the direction of displacement from the location of the railway station on the 1 : 50 000 scale map.

potentially be converted into a park on the criteria adopted. The areas of these two zones are shown in Table 6.9. Zone 2 progressively increased in area with the reduction of scale. From the 1 : 50 000 scale map to the 1 : 250 000 scale, zone 2 increased by 8 per cent in area, while for the 1 : 625 000 scale map there was approximately twice as much increase in area (18 per cent). Zone 1, however, did not show the same pattern of area change. Between the 1 : 50 000 and the 1 : 250 000 scale maps, zone 1 suffered a large decrease in area (equal to 26 per cent), and this was followed by a very large area increase of 114 per cent on the 1 : 625 000 scale map. In total, between the first two scales (i.e. 1 : 50 000 and 1 : 250 000), the total area of the two zones decreased by 3 per cent; and between the 1 : 50 000 and the 1 : 625 000 scale maps, the area increased by 51 per cent. In other words, depending on the scale used for the GIS analysis, the total land area of the parks could vary by more than half.

It can be seen from Figures 6.9–6.11 that, in addition to the two zones that were suitable for conversion into a park, the overlay generated a series of smaller zones (with an area less than 500 000 m^2), which changed in area, shape and number for the three scales. These were unique to the British study area. The number of these smaller zones decreased with the reduction of scale. In terms of their total area, there was a 70 per cent increase between the 1 : 50 000 and 1 : 250 000 scale maps, while between the 1 : 50 000 and the 1 : 625 000 scale maps there was a 45 per cent decrease in area.

Within the British study area, the relevant features used in the overlay (i.e. common to all three scales) were as follows: two railways, two railway stations, one river, the outline of the city of Canterbury and 13 roads. It was the combined generalisation of all of these features that contributed to the changes in the number, shape and area of the different zones for each scale used. In section 5.4.1, it was shown that the roads and the rivers were displaced the most at both scales and had the most length change. Therefore, they played the largest role in affecting the results from the overlay operation. In contrast, the railways contributed the least because they suffered less displacement and had the least change in length.

The outline of the city of Canterbury was the only urban area that was present across the three scales. Changes in the urban outline of the city of Canterbury in terms of its shape, position and total area all influenced the

Table 6.9 The areas of the zones that resulted from an overlay to determine the potential location of a new park for the region of Canterbury and East Kent.

Areas of the different zones per scale (m^2)					
1 : 50 000		1 : 250 000		1 : 625 000	
Zone 1	712 976	Zone 1	529 378	Zone 1	1 526 598
Zone 2	1 379 777	Zone 2	1 491 389	Zone 2	1 635 338
Total	2 092 753	Total	2 020 767	Total	3 161 936

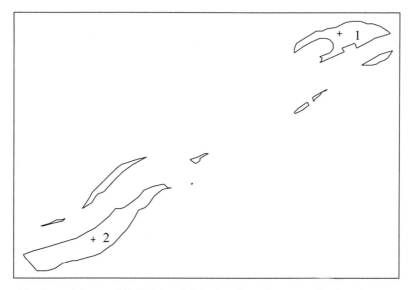

Figure 6.9   Potential areas (labelled 1 and 2) for location of a new park using the 1 : 50 000 scale map for the region of Canterbury and East Kent (João *et al.*, 1993).

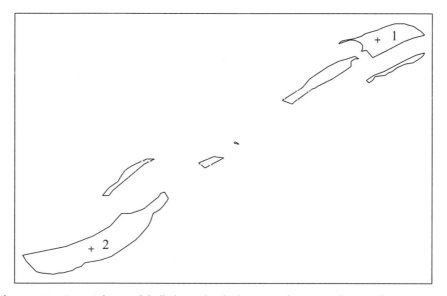

Figure 6.10   Potential areas (labelled 1 and 2) for location of a new park using the 1 : 250 000 scale map for the region of Canterbury and East Kent (João *et al.*, 1993).

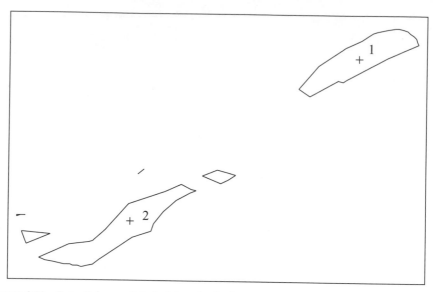

Figure 6.11   Potential areas (labelled 1 and 2) for location of a new park using the
1 : 625 000 scale map for the region of Canterbury and East Kent (João *et al.*, 1993).

results of the overlay. In Figure 6.12 it is shown how the shape and position of
the outline of the city of Canterbury altered across the different scales.

In Figure 6.12, the outline of the city of Canterbury at the scale of 1 : 50 000
is contrasted with its representation at the two smaller scales. The area
occupied by the urban outline of the city at the scale of 1 : 50 000 was
10 445 332 m². This area decreased to 9 212 878 m² for the 1 : 250 000 scale
map, and decreased even further, to 7 747 400 m², for the highly generalised
outline represented on the 1 : 625 000 scale map.

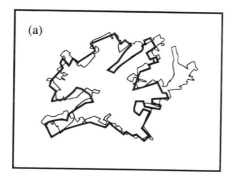

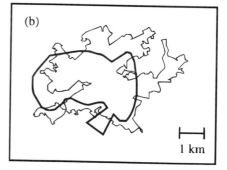

Figure 6.12   The urban outline of the city of Canterbury. (a) The outline of the urban area at
scale 1 : 50 000 versus 1 : 200 000. (b) The outline of the urban area at scale 1 : 50 000 versus
1 : 625 000. In both cases the 1 : 50 000 scale is represented with a thinner line. (Urban
boundaries taken from OS digital data.)

The large change in area of zone 1 (see Figures 6.9–6.11) was mainly influenced by the change in the urban outline of the city of Canterbury. For the 1 : 250 000 scale map, a separate 'corridor' appeared south-west of zone 1, caused by a gap that appeared in the outline of the city in the north-east. On the 1 : 500 000 scale map, the outline of Canterbury shrank even further, and as a consequence zone 1 became enlarged to include this corridor. The shape and size of zone 2 was mainly influenced by the position of the A28 road relative to the Great Stour river. As they moved further apart, zone 2 became enlarged.

The changes in the position of the two railway stations also influenced the results. In Figure 6.13 are shown the directions and the magnitudes of the displacements of the Canterbury East and West railway stations between the 1 : 50 000 and the two smaller-scale maps. These two stations were the only railway stations common across the three scales. The displacement of the West Station between the 1 : 50 000 and 1 : 250 000 scale maps (80 m) and between the 1 : 50 000 and 1 : 625 000 scale maps (180 m) corresponded to the same amount in terms of map units; that is, 0.3 mm. The displacement of the East

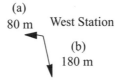

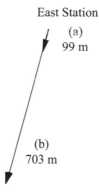

Figure 6.13   The displacement of the two British railway stations within Canterbury: the West Station and the East Station. The vectors indicate the direction of displacement from the location of the railway stations on the 1 : 50 000 scale map. (a) The displacement between the 1 : 50 000 and the 1 : 250 000 scale maps. (b) The displacement between the 1 : 50 000 and the 1 : 625 000 scale maps.

Station between the 1 : 50 000 and 1 : 250 000 scale maps corresponded to 0.4 mm in map units, while the displacement between the 1 : 50 000 and 1 : 625 000 scale maps (in the same direction as before) corresponded to a shift of 1.1 mm on the map.

The position of the railway stations was closely associated with the position of the railways themselves (as even in the most generalised maps the *relative accuracy* of features is usually carefully maintained by the cartographer). This meant that when the railways were displaced, inevitably the railway stations were displaced as well. The very large displacement of the Canterbury East Station occurred on the 1 : 625 000 scale map because this station was located on the section of the railway that suffered the most displacement at this scale (due to the enlargement of the circular road around Canterbury, as discussed in section 5.4.1). In terms of the overlay analysis, these displacements meant that the centroid of the two circles of 5 km radius shifted, sometimes in different directions, from 80 to 703 m, encompassing different features or portions of features as a consequence. The relative position of the two railway stations also played a role in affecting the results. As the railway stations moved further apart, there was less overlap and the total area of the railway station buffer increased. At 1 : 50 000 the combined buffer area was approximately 89 km^2, increasing to 90 km^2 at 1 : 250 000 and finally to 94 km^2 for the 1 : 625 000 map.

By comparing the British with the Portuguese results, it appears that the British data varied more across scales. The potential sites for the British park changed more in shape, position and area. One reason for this was that in the Portuguese case neither the urban area nor the railway line influenced the siting of the park. This meant that there were two fewer factors that could introduce variability into the Portuguese results. Another difference was that there was only one railway station in the Portuguese study area. Although it changed position slightly, the railway station buffer remained constant in area (78 km^2) across all three scales. In contrast, Canterbury had two railway stations common to the three scales, and because they were relatively close together, the 5 km buffer of the stations partially overlapped. This meant that not only did the position of the railway stations change, but there was also a change in the total area of the resulting buffer. It was a combination of these factors that meant that the results of the GIS overlay operation using the British data were affected more drastically by generalisation compared with the Portuguese results.

6.2.2   Overlay operation using all the features present in the maps

The overlay operation was repeated using all of the features present in the maps. This was to simulate the overall impact of the generalisation process on a typical GIS analysis, and also to evaluate the more probable result that a

GIS user would have obtained (assuming that all of the features present in the maps were used). In this section only the data for the two largest scales are shown (1 : 50 000 and 1 : 250 000 for the British study area; and 1 : 50 000 and 1 : 200 000 for the Portuguese). The results for the smallest-scale maps are the same as shown in section 6.2.1. The results of the overlay using all of the features present in the Portuguese study area are shown in Table 6.10 and Figures 6.14 and 6.15.

For the Portuguese study area, at the scales of 1 : 50 000 and 1 : 200 000, there was the same *number* of zones (six in total) with an area greater than 500 000 m^2 that could be considered suitable for conversion into a park (see Table 6.10). However, these six zones (numbered in Figures 6.14 and 6.15) changed across the two scales. For example, zone 78 was unique to 1 : 50 000 and zone 86 was unique to 1 : 200 000. For the scale of 1 : 500 000, the six areas were reduced to four, as shown in Table 6.8 and Figure 6.7. The total area of the zones suitable for conversion into a park also decreased with scale. From 1 : 50 000 to 1 : 200 000 there was a 41 per cent reduction in area, and between the 1 : 50 000 and the 1 : 500 000 scale maps there was an 81 per cent reduction.

The differences in these results (compared with the overlay carried out using only the features that were present across all the scales) were due to generalisation of individual features in addition to the high variability of the number of features present in the maps of different scales. The area or length of the different feature types relevant to the overlay operation for the three scales are shown in Table 6.11. These differences in total area and length had serious consequences on the overlay results, as seen when comparing Figures 6.7, 6.14 and 6.15.

The urban areas also affected the results, although this influence was relatively small. Most of the houses were within 100 m of a road and were there-

Table 6.10   The areas of the zones that resulted from the overlay operation using all the features present in the Portuguese study area.

Areas of the different zones per scale (m^2)			
1 : 50 000 scale map		1 : 200 000 scale map	
Zone 15	4 399 543	Zone 15	4 249 413
Zone 18	5 685 172	Zone 39	1 983 617
Zone 28	757 957	Zone 54	954 778
Zone 50	2 622 748	Zone 50	826 333
Zone 78	1 083 181	Zone 86	671 802
Zone 115	1 350 683	Zone 115	708 358
Total	15 899 284	Total	9 394 301

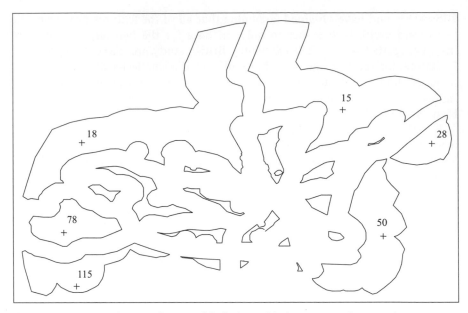

Figure 6.14   Potential zones (the areas labelled) suitable for a new park, using the 1 : 50 000 scale Portuguese map and all the features present.

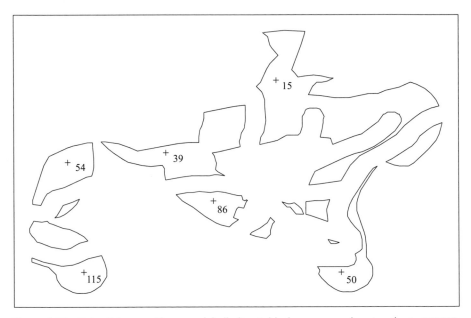

Figure 6.15   Potential zones (the areas labelled) suitable for a new park, using the 1 : 200 000 scale Portuguese map and all the features present.

Table 6.11   The length or area of all of the features used in the overlay for the three Portuguese scales (the number of features is in brackets).

Feature type	1 : 50 000 scale map	1 : 200 000 scale map	1 : 500 000 scale map[a]
Rivers	240 756 m (185)	79 759 m (19)	21 398 m (2)
Roads (excluding footpaths)	70 819 m (41)	46 063 m (11)	32 579 m (4)
Railway	14 154 m (1)	13 846 m (1)	13 518 m (1)
Railway station's buffer[b]	78 km^2 (1)	78 km^2 (1)	78 km^2 (1)
Urban areas[c]	1 418 595 m^2 (127 houses)	1 068 478 m^2 (31 houses)	—[d]

[a] The number of features at this scale is the same as at the scales of 1 : 50 000 and 1 : 200 000 for the features used in the overlay operation described in section 6.2.1.
[b] The area of the railway station corresponds to the area occupied by the 5 km buffer.
[c] The urban areas correspond to the area of the houses plus a 50 m buffer around them.
[d] The town of Fronteira is only represented by a circle symbol at the scale of 1 : 500 000.

fore located within discarded zones anyway. In Figures 6.16 and 6.17 are shown the houses present in the Portuguese study area, which were used in the overlay operation at the scales of 1 : 50 000 and 1 : 200 000 respectively. The houses in the figures have 50 m buffers drawn around them so as to simulate the outline of urban areas.

For the Portuguese maps, at the scales of 1 : 50 000 and 1 : 200 000, as well as the six zones that were suitable for conversion into a park, the overlay

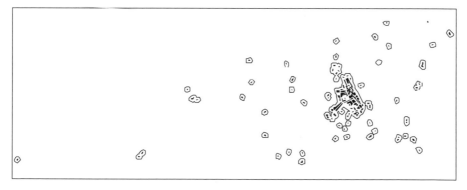

Figure 6.16   The houses of the Portuguese study area and the surrounding 50 m buffers used in the overlay analysis, taken from the scale of 1 : 50 000.

Figure 6.17   The houses of the Portuguese study area and the surrounding 50 m buffers used in the overlay analysis, taken from the scale of 1 : 200 000.

generated a series of smaller zones (less than 500 000 m^2) which changed in area, shape and number across the different scales (see Figures 6.14 and 6.15). The number of these areas decreased with the reduction of scale (a 56 per cent reduction from 1 : 50 000 to 1 : 200 000). The total area of these smaller zones also decreased with the reduction of scale, although not so severely (a 28 per cent reduction between the 1 : 50 000 and the 1 : 200 000 scale maps). This indicates that although there were fewer of these small zones, they were larger in size. In summary, for the Portuguese overlay operation, what influenced the results the most was the presence or absence of roads and rivers at each scale. For example, zone number 86 is unique to the scale of 1 : 200 000 because some roads disappear at this scale, therefore making it a viable area for conversion into a park. In contrast, zone 50 was much reduced in area at the scale of 1 : 200 000 because of the elimination of certain rivers.

When the overlay operation was carried out using all of the features present in the British maps, the number of areas that were suitable for conversion into a park changed for each scale. On the 1 : 50 000 scale map there were 13 zones, each with an area equal to or greater than 500 000 m^2. These were all therefore suitable for conversion into a park, according to the initial specifications. For the scale of 1 : 250 000, the number of these zones decreased to 11. The data for both of these scales can be seen in Table 6.12. A further reduction to two zones was obtained when using all of the features present on the 1 : 625 000 scale map (see Table 6.9). The two potential zones that were generated when using the 1 : 625 000 scale map can be seen in Figure 6.11, and the different

Table 6.12 The areas of the zones that resulted from the overlay operation using all of the features present in the British maps.

Area of the different zones per scale (m^2)			
1 : 50 000 scale map		1 : 250 000 scale map	
Zone 6	764 589	Zone 6	515 320
Zone 7	4 544 142	Zone 7	2 159 915
Zone 8	965 659	Zone 8	520 147
Zone 11	2 114 804	Zone 11	1 773 442
Zone 14	1 767 389		
Zone 15	539 823	Zone 15	620 584
Zone 41	762 088	Zone 41	586 695
Zone 49	919 632		
Zone 57	1 739 215	Zone 57	1 246 506
Zone 148	2 899 160	Zone 148	1 762 313
Zone 505	944 287	Zone 505	710 332
Zone 518	1 741 056	Zone 518	1 451 092
Zone 532	893 699	Zone 532	1 112 531
Total	20 595 543	Total	12 458 877

suitable zones generated by using the data of the 1 : 50 000 and 1 : 250 000 scale maps can be seen in Figures 6.18 and 6.19 respectively. In these diagrams, the potential parks are numbered.

The total area of the potential zones that could be converted into parks decreased with scale. Between 1 : 50 000 and 1 : 250 000 there was a 39 per cent decrease in total area, while for the scale of 1 : 625 000 an 85 per cent decrease was observed. As with the results of the Portuguese overlay, in addition to the zones that were suitable for conversion into a park, the British overlay generated a series of smaller zones that changed in area, shape and number for the three scales (compare Figures 6.11, 6.18 and 6.19).

The road buffers were the main feature that was responsible for the breaking up of the study area and consequently the generation of so many different small zones (i.e. with an area less than 500 000 m^2). Many of the urban areas lay within the 100 m road buffers and hence played a lesser role. The one exception was the large urban outline of the city of Canterbury. The total number of individual areas decreased with the reduction of scale (a 40 per cent reduction from 1 : 50 000 to 1 : 250 000, and a 92 per cent reduction between the 1 : 50 000 and the 1 : 625 000 scale maps). The total area of these smaller zones also decreased with the reduction of scale (a 42 per cent reduction between the 1 : 50 000 and the 1 : 250 000 scale maps, and a 94 per cent reduction for the scale 1 : 625 000). The length and area of the different feature types relevant for the overlay operation for the three scales when all the features present in the British maps were used are shown in Table 6.13.

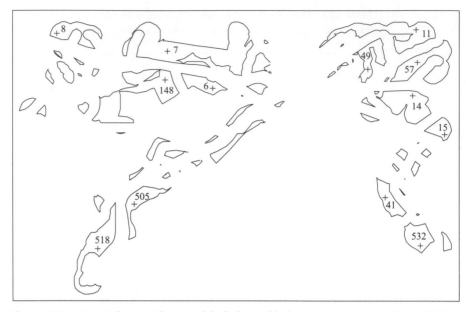

Figure 6.18 Potential zones (the areas labelled) suitable for a new park, using the 1 : 50 000 scale British map and all the features present.

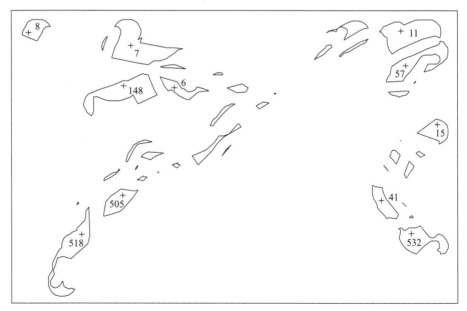

Figure 6.19 Potential zones (the areas labelled) suitable for a new park, using the 1 : 250 000 scale British map and all the features present.

Table 6.13 The length or area of all of the features used in the overlay for the three British scales (the number of features is in brackets).

Feature type	1 : 50 000 scale map	1 : 250 000 scale map	1 : 625 000 scale map[a]
Rivers[b]	79 296 m	42 086 m	16 180 m
	(85)	(15)	(1)
Roads[b]	337 409 m[c]	325 490 m[d]	81 199 m[e]
	(646)	(512)	(13)
Railways	34 739 m	34 770 m	35 209 m
	(2)	(2)	(2)
Railway stations'	245 329 400 m²	244 298 100 m²	94 166 944 m²
buffers[f]	(6)	(6)	(2)
Urban areas	17 641 300 m²	21 393 700 m²	7 747 400 m²
	(66)	(34)	(1)

[a] The number of features at this scale is the same as at the scales of 1 : 50 000 and 1 : 250 000 for the features used in the previous overlay operation described in section 6.2.1.
[b] As explained earlier, the roads and rivers in the 1 : 50 000 and 1 : 250 000 scale maps are very fragmented; therefore the numbers in brackets really refer to the number of *segments* of features.
[c] Excluding footpaths, bridleways, roads used as rights of way, and other non-urban roads.
[d] Excluding long-distance footpaths.
[e] Excluding minor roads.
[f] The area of the railway stations corresponds to the area occupied by the 5 km buffers.

In the British case, between the 1 : 50 000 and 1 : 250 000 scale maps, the river buffers were the most critical in affecting the results. Zone 7, for example, had the largest decrease in area at the scale of 1 : 250 000, because a river was eliminated on the eastern side of the zone. For a similar reason, zone 14 completely disappeared at this same scale, because rivers were eliminated and the zone disappeared with them. Zone 532 at the scale of 1 : 250 000 merged with a small neighbouring area and was therefore enlarged: this was because a road was eliminated between the two areas. In contrast, zone 505 decreased in area when it lost a thin long section due to the displacement of the railway relative to the river.

The number of features retained at smaller scales heavily influenced the area and length occupied by the different feature types for both study areas. Depending on the feature type, and according to the overlay constraints, an increase in length or area could either cause a decrease or an increase in the total area suitable for a new park. For example, if the number of railway stations decreased (as in the British data) or if there was a large decrease in the length of rivers (as in the Portuguese case), then this would lead to a smaller number or a decrease in area for the potential sites. On the other hand, a decrease in the area occupied by urban zones would increase the area that could be converted into a park. Therefore, the elimination of features became more critical than the displacement or simplification of individual features.

For both countries, the presence and absence of rivers caused the most changes to the results, as this was the feature type that was eliminated most as the scale of the map decreased. Using all the features for the overlay operation led to much greater variation across scales and more accurately emulated what would happen with a typical GIS manipulation. However, it is interesting to note that the *decrease* in area of the potential park zones for the Portuguese and British middle-scale maps was actually very similar (41 per cent and 39 per cent respectively). The same was true for the smallest-scale maps (81 per cent for the Portuguese parks and 85 per cent for the British parks).

Both the point-in-polygon and the overlay operation showed that generalisation can have an important effect on GIS map manipulations. This effect can be large, although it is often variable. With the point-in-polygon operation, the observed effect is increased if it is necessary to take into account *which* points fell within the polygons. This is the case even when only one set of features (i.e. the roads) are being generalised. For the overlay operation, the generalisation effects were exacerbated by the use of many different types of features. In the final chapter, the relevance of these results is discussed, and new ways of minimising generalisation effects within a GIS are proposed.

CHAPTER SEVEN

The importance of generalisation effects: repercussions for the future of GIS

While scientists have long been aware that generalisation has an important influence on map features and analysis, surprisingly little research has been devoted to quantifying it. This study has proposed a methodology for measuring generalisation effects (by separating out generalisation from other sources of map error) and has revealed precisely how large they can be. It has investigated the relative importance of displacement, change in line length and elimination of features. It is concluded that generalisation is a major contributor to the variation in the map features as the scale is changed, and that often large inaccuracies are caused by generalisation. Moreover, it is also shown that these effects can have important consequences on the results of typical GIS map manipulations. Depending on the scale used, different results are obtained from a point-in-polygon and an overlay operation.

This chapter suggests ways of dealing with generalisation within a GIS. Clearly, wherever appropriate, maps of a large enough scale should be used, although they are not always available. A more fundamental approach is proposed, based on the findings that automatic generalisation causes fewer effects than manual generalisation. It is suggested that automatic methods should be used to a greater extent, for both model and cartographic generalisation, in order to control for generalisation effects. An approach to generalisation within a GIS is described which takes this into account while also giving considerable flexibility to the user.

7.1   A summary of the findings on quantifying generalisation effects

The measurement of displacement was the most powerful indicator of carto-graphic generalisation for both study areas. In absolute ground distance terms, for all of the features of both countries, the displacement of lines always increased with decreasing scale. Also, this displacement, for the middle- and the smallest-scale maps of both countries, had the largest absolute change in comparison to other measures. Displacement not only led to lower (absolute) positional accuracy, but it also changed the relative position of the features (or portions of them). As a consequence, the lateral shift of a line was the most important factor in introducing spatial error due to generalisation when carry-ing out GIS map manipulations. All feature classes of both countries showed *maximum* displacements greater than the 0.4 mm error cut-off point (determined by analysing the sources of random positional error). For example, the maximum displacement for both the British and the Portuguese roads, at the scales of 1 : 250 000 and 1 : 200 000, was 1.3 mm. This is over three times the value expected if only random errors were to affect the posi-tional accuracy of the features. It is interesting to note that this maximum displacement of the Portuguese and the British roads was nearly twice the value typically allowed for the 1 : 250 000 Australian topographic database (Australian Survey, 1992).

Although it is perhaps not surprising that certain features had sections that were excessively displaced, it was less expected that for some of the features the *average* displacement exceeded the value of the cut-off point. The roads and railways for the Portuguese 1 : 500 000 scale map, and the boundaries for the OS 1 : 625 000 scale map, were all considerably displaced. Moreover, intersec-tion points, commonly assumed to be the most accurately positioned of fea-tures, often showed large displacements. In fact, only one of the intersection points on the OS 1 : 625 000 scale map was positioned within 0.4 mm of its location on the 1 : 50 000 scale map. That is not to say that displacements less than the 0.4 mm cut-off point were not also influenced by generalisation, but for small displacements it was impossible to dissociate generalisation from random errors.

For displacements larger than the 0.4 mm cut-off point, there was the possi-bility of the occasional occurrence of large gross errors. In the case of one of the Portuguese roads and the British boundaries, it seemed likely that there were some gross errors affecting the positional accuracy of the features. This distinction between *conscious* generalisation and *unconscious* gross error could only be made by analysing the cartographic reason or need for generalisation. When very large displacements could not be explained by the need to avoid other features, or by the need to simplify the features' detail, then the existence of an unconscious gross error was assumed. This compounds the difficulty of distinguishing between generalisation and other error. Unless the information related to *when* and *how* features are generalised is known, it is difficult to ascertain whether the inaccuracy of a feature has been caused by generalisation or other sources of error. This makes it extremely difficult to reverse-engineer generalisation and identify generalisation effects in data quality statistics.

Two types of displacement were identified, termed *local* and *global* displacement. With local displacement, particular sections of a line might lose detail (e.g. a particular road bend) when simplified and, as a consequence, local sections of a feature might be displaced sideways. Global displacement, as the name indicates, affects a large portion of a feature – or even, in certain circumstances, the whole feature. In other words, the feature on a particular map sheet might be completely displaced in order to avoid other features. Local displacement will often be related to a loss of angularity and a loss of length. However, a feature that suffers global displacement can still retain most of its initial length and angularity. For example, on the Portuguese maps the roads and railway suffered more displacement at the scale of 1 : 500 000 than the boundary and rivers, but it was the boundary and rivers that had the greatest length change at this scale. The lack of change in length and angularity might fail to highlight the generalisation of a feature, another reason why displacement was found to be a more powerful indicator of generalisation.

In the case of length and angularity measurements, this study uncovered some counter-intuitive results. The length and angularity of a feature generally, but not invariably, reduced as scale decreased. Occasionally, they actually increased with the reduction of scale. This was caused by lines being displaced or exaggerated, such as in the case of the railway south of Canterbury. This contradicts what certain cartographic theorists have predicted in the past. For example, Maling (1989, p. 67) reported on the 'phenomenon that the larger is the scale of the map, the greater is the length of the measured line'. While this is certainly a trend, it is not inviolate as a principle.

However, these counter-intuitive results were mainly observed when individual features were considered. When they were grouped by feature type they showed less variation in the measurements of length and angularity. One reason for this is that, when generalising, the cartographer tries to maintain the *topology* of the features as accurately as possible; in other words, the cartographer might sacrifice absolute accuracy (the position of the features relative to a coordinate system) in order to maintain relative accuracy (the position of the features relative to each other). It happens, therefore, that roads can be exaggerated and have a large increase in length (such as the enlargement of the circular road around Canterbury) and as a consequence contribute to a large decrease in length of another road (such as the link road between the circular road and the A2 dual carriageway). In other words, the changes in length of different features balance each other out. However, it was not expected that, in the case of the 1 : 625 000 scale British map, the railway *feature class* would increase in length. Both railways showed an increase in length at the scale of 1 : 625 000, albeit slightly. One of the reasons for such an effect was that both of the railways had bends that were exaggerated and made sharper. The railway to Dover, for example, had to be displaced around the enlarged circular road that surrounds the city of Canterbury. The displacement of a portion of the railway in that particular direction caused the increase in overall length.

The results were found to be very variable according to feature type, and especially according to individual features. It was therefore often difficult to determine beforehand which of the feature classes (e.g. road or railway) would be most affected by generalisation and for which of the scales. For both countries, the feature class that presented the largest lateral shift at one scale was not necessarily the same as that having the largest value at another scale. This makes it problematic to generate hard and fast rules for individual feature classes, and shows the importance of concentrating on the geometric characteristics of a line rather than what it represents (e.g. a river versus a road). The length and angularity of the rivers of both countries, for example, tended to be more affected than roads and railways, but this is not necessarily applicable to rivers in general: it could be because of the individual characteristics of the rivers that were present in the particular areas studied. A more 'wiggly' meandering river on one map, for example, will have very different generalisation effects compared with a straight river. Even within the same feature type, not all features suffered the same generalisation in terms of local and global displacement. It was therefore difficult to predict exactly which feature type would cause, for example, more of an impact on the results of the overlay operation. In the end, in order to determine which features influenced the variation in the results the most, it was necessary to identify the distribution of the individual features in relation to the results of the overlay. The variability between features can be illuminated by the following quote (Müller *et al.*, 1995, p. 9):

> The landscape of geographical features portrayed on topographic maps, for example, can vary almost to infinity. This great variation creates generalisation problems which cannot all be foreseen and the research required to cover all cases is so complex and so demanding that it would not be economical.

This statement can also apply to the variability between countries. On average, most Portuguese features had more changes in length than the British ones. In addition, Portuguese features suffered more lateral shifts than the British features at the smallest scale (1 : 500 000 for Portugal and 1 : 625 000 for Britain). For the smallest scale, the Portuguese features had larger maximum displacements and on average the total areal displacement was also higher (with two feature classes having average displacements larger than the 0.4 mm cut-off point). In contrast to this, in the case of the intermediate scales (1 : 200 000 for Portugal and 1 : 250 000 for Britain), the Portuguese features suffered less lateral shift on average than the British features. For the intermediate scale, the total areal displacement was usually higher for the British features and the mean size of the polygons generated was also larger. However, none of the features of both countries at the intermediate scale had an average sideways shift larger than the 0.4 mm cut-off point. On the whole, it seems that the Portuguese features suffered more serious generalisation effects. However, without analysing further study areas, the results remain indicative and it is not valid to extrapolate to other areas of the same country, let alone to different countries.

Despite the fact that there was considerable inconsistency between the two countries and between features, one general pattern did emerge. This was that the more detailed a feature (measured in terms of angularity) on a larger-scale map, the more likely it was to have a very high reduction of that angularity. In other words, the more 'wiggly' a line, the more vulnerable it was to generalisation and the more it was 'straightened out'. A correlation was found between the mean angularity of each feature type at the scale of 1 : 50 000 with the change in angularity suffered by each of those feature types at the smallest scale ($r^2 = 0.96$, $P < 0.05$). This suggests an ability to predict the amount of generalisation from a given initial angularity.

The generalisation of the different features of the two study areas seemed to depend on the initial state. Not only did the initial amount of angularity influence how much angularity was lost at smaller scales, but the initial number of features also determined how many were eliminated. It was found that the greater the quantity of roads and rivers was at the largest scale (measured in terms of total line length), the more features would be eliminated. For example, in the Portuguese study area, initially with three times more rivers than roads, at the scale of 1 : 500 000 nearly twice as many rivers than roads were eliminated in terms of total length. In the British case, initially with 11 times more roads than rivers, the opposite happened, and almost twice as many roads than rivers were eliminated in terms of total length at the scale of 1 : 625 000.

However, the elimination of the *number* of features with decreasing scale, in the case of the Portuguese study area, did not decrease according to Töpfer's law. While Töpfer and Pillewizer acknowledged that their law would not always hold exactly (such as in the case of the Swiss cartographer who retained many more lakes than expected because 'the compiler did not dare to omit certain well known lakes in Switzerland'; Töpfer and Pillewizer, 1966, p. 15), the discrepancies between the actual and the expected number of features for the Portuguese maps were much larger than the law allowed for. This may be partly due to the maps in this study being at a much larger scale than the atlas maps used by Töpfer. Any hard and fast rule in connection with generalisation must be treated with a certain amount of scepticism.

The research work in this book was carried out to confirm the assumption that generalisation effects are important to GIS users and cartographers alike. While it was shown that generalisation effects are sizeable in relation to other errors introduced by the cartographic process, it was also necessary to show that the effects are large enough to impinge upon a GIS analysis. In other words, effects have to be large enough to throw into doubt the conclusions from a 'typical' GIS map manipulation. If generalisation effects do significantly influence the results from a GIS map manipulation, then obviously they cannot be ignored. The findings of this research demonstrated this to be true. The point-in-polygon and the two overlay operations, using standard data sets, revealed a large variation in the results depending on the source scale (and hence the amount of generalisation) used.

The different types of generalisation effects that features might have suffered (such as displacement and length change) are compounded when features are used in a GIS map manipulation. As a consequence of this, the changes in the results of any one GIS map manipulation due to generalisation will be potentially even more unpredictable than the changes to individual features. For example, with the point-in-polygon operation, the generalisation effect of length change (which caused the area of the road buffers to vary) and the displacement of the many individual roads were combined. This 'combined action' caused the total number of points that fell within the road buffers to fluctuate. Occasionally, when points fell in and out of the road buffers, the results remained the same across the different scales: the generalisation effects were apparently cancelling each other out. However, when the specific identity of the points was taken into account, the results varied more widely, depending on the scale used. This time, as the scale decreased, the number of specified points that fell within the road buffers decreased by more than half for both countries.

For most GIS map manipulation, any generalisation effect is also compounded when two or more different types of features are considered simultaneously. The generalisation of individual features acting in unison was particularly striking in the case of the overlay operation. As the relative position of the features was altered by generalisation (by loss of detail and by displacement), the resulting zones that could be converted into a park varied considerably. Even when individual features show small generalisation effects, the cumulative impact on a GIS analysis that uses several different generalised features in combination can be much larger (when, for example, two features have small displacements in opposite directions). As more features were used in the overlay operation, the changes in the results for the total area, position, number and shape of the resulting parks were larger, and less predictable.

Although it is possible that not all GIS map manipulations would be so severely affected by generalisation, it is a reasonable inference that *all* users of GIS should be aware of the problems of generalisation when carrying out an analysis, and that they should usually seek to work with maps of the largest available scale. While it is unlikely that a geographer would deal with a 100 m road buffer using a map at a scale of 1 : 625 000, it is not inconceivable that this might occasionally be necessary if the study was covering a particularly large region or if larger-scale maps were unavailable. Indeed, it is possible that generalisation problems may often be much larger than the variation demonstrated in this study: the use of base maps from multiple different map-producing agencies could lead to worse results than those originating from within a single family of maps.

The results might well have differed if other study areas had been used. Larger or different (e.g. more mountainous) study areas could have generated either larger or smaller generalisation effects. Some of the inconsistencies, between countries and between features, may be partly due to the small numbers of certain features. By taking a larger sample size, it might have been possible to extract more overriding patterns. A large effort was involved in digitising, editing and preparing the data for measurement, despite the use of an efficient and sophisticated digitising technique (i.e. VTRAK), and the use of GIS to carry out the measurements automatically. The effort in preparing data imposed conditions on the number and size of the study areas possible: two study areas, each with an approximate area of 200 km^2, were used. Although the study areas were of a reasonable size, the number of features that were simultaneously present at all scales was limited (especially due to the elimination of features on the smaller-scale maps). This meant that the study was restricted in scope with regard to both the number of different feature classes and the number of features per class that were measured. Because of this, the results of this study were very much indicative rather than definitive.

Without carrying out further studies, it will not be known whether larger study areas of different regions will lead to an increase or a decrease in the variability of the generalisation effects for different features. It is possible that more patterns will begin to emerge, but probably, and in accordance with Müller et al. (1995), the variation of the generalisation of features is virtually infinite. Even when an overall trend is found for certain groups of features (e.g. rivers or roads), some individual features will always be an exception to the rule, as was found in this study. Because any generalisation rule, if introduced automatically, has to be applied to an individual feature, rather than to an 'average' feature, problems will invariably arise. The implications of this are that it is not possible to reverse-engineer generalisation with confidence. Therefore, in order to control and quantify generalisation effects when using a GIS, it will be necessary to develop an automated system that would generalise while simultaneously maintaining close control of the consequences of generalisation. Such a system is described in section 7.3. This type of system would require automatic algorithms to generalise the data. Therefore, this study also compared the generalisation effects of features that resulted from manual generalisation with features that were generalised using a typical line simplification algorithm.

7.2 A comparison of effects of manual and automated generalisation

This study has presented quantitative evidence that generalisation effects are typically greater in manually digitised paper maps than in those produced by the Douglas–Peucker algorithm. Automatic generalisation using this algorithm retained length and angularity very well and, most importantly, displaced features much less. In other words, it caused much less distortion than manual generalisation. This is because the Douglas–Peucker algorithm only filters the high-frequency components, causing a reduction in the local detail of the line without the global displacement from which the manually generalised lines often suffered. This is related to the way in which most line simplification algorithms work – they do not displace a line along its entire length, as even if the line loses the majority of its points, its first and last point will remain constant.

In the worst possible case of displacement caused by line simplification algorithms, a feature becomes a straight line. Müller (1987), for example, evaluated eight different line simplification algorithms by comparing the displacement caused by the algorithms against this worst case. (Müller called this the 'Standard Measure of Displacement'.) In the case of simplification algorithms, the line features only suffer local displacement and do not suffer global-type displacement. In an automated environment, global displacement would have to be carried out by the still underdeveloped displacement algorithms (see section 1.3.1). Different line simplification algorithms cause varying amounts of local displacement (McMaster, 1987; Müller, 1987) but the displacement caused by manual generalisation is almost always larger, because of the extra sideways shift caused by global displacement. These findings support the results of Beard (1988), who found that, for a thematic map depicting land and water areas, both positional and attribute errors were reduced by automatic generalisation.

The fact that the generalisation effects caused by the Douglas–Peucker algorithm were less than those caused by manual generalisation is a reflection not only of the use of different procedures (automatic versus manual) but also of generalising for different *purposes*. It can be argued (especially for such small tolerances as the ones used) that the Douglas–Peucker algorithm is a data reduction technique that is more appropriate for model generalisation (for analysis purposes) rather than for cartographic generalisation (for display purposes). In contrast, the maps used in this study had been generalised for cartographic purposes. As discussed in section 2.2, both cartographic and model generalisation introduce error into generalised data sets, but the errors resulting from model generalisation will normally be less, because the data reduction is carried out under quantitative-statistical control. Cartographic generalisation will displace geographical features along their entire length when there is conflict between features or will cut across large loops. It was the lack of global displacement and a reduction of local displacement that caused the lines resulting from automated model generalisation to have fewer generalisation effects than those resulting from manual cartographic generalisation. As a consequence of the small displacement, the automatically generalised lines also suffered less change in line length and angularity.

In statistical terms, automated generalisation appeared to produce better results but, graphically, the appearance of the line may be inferior. For very large tolerances, the Douglas–Peucker algorithm is known to cause results that are considered to be inadequate visually (Visvalingam and Whyatt, 1990). In practice, GIS analyses are likely to be more reliable when using lines that resulted from automated generalisation. However, at present, if any sort of pictorial analysis is needed, then maps derived from manual generalisation would often be visually clearer. The development of new algorithms, such as the one by Li and Openshaw (1993), which creates generalised features that resemble those resulting from manual generalisation, is a step towards achieving acceptable graphical results, although the problem of automated displacement remains largely unsolved.

With the development of new algorithms that minimise generalisation effects, or the improvement of existing ones, it is likely that the use of more automated generalisation methods (e.g. in national mapping agencies) would help to reduce generalisation effects. This is supported by the fact that, frequently, digital maps do not need to be as crowded as traditional analogue maps. Paper maps have an abundance of symbols, and the generalisation of the maps also needs to take into account the space occupied by the different labels and point and line symbols. Minimisation of the number of these labels and symbols (since in spatial databases they are not required to be drawn on the map but, instead, can be stored in the database and called up when required) might help to reduce the clutter of maps and therefore lessen the amount of generalisation needed for the different features. Phillips and Noyes (1982), for example, reported a study to assess the reduction of visual clutter on a 1 : 50 000 scale geological map by removing topographic symbols that were only of minor importance to the map reader. They concluded that this reduction of clutter significantly improved the clarity of the maps.

At a more theoretical level, the generalisation of features could be made more sophisticated. Instead of generalising maps on the basis of display or communication criteria (the 'what looks right' factor), maps for specific purposes (e.g. hydrological maps) could be generalised, with the theme of the map as the overriding criterion. Following up the hydrological example, lakes should be generalised by also taking into account their depth, and not just their area, shape and spatial arrangement (Beard, 1991a). Liebscher (1982) argued that there was insufficient communication between cartographers and hydrologists in the production of the European CERCO digital maps. According to the author, criteria such as flow and water quality should be used when generalising, rather than the more usual arbitrary, aesthetic choices. The main criterion used by cartographers when generalising rivers is the width of the river feature (see Table 1.2). However, this does not always coincide with the flow rate. Catlow and Daosheng (1984) suggested that drainage network data, previously collected with map production strictly in mind, should be given a 'structure' based on the stream ordering.

Gardiner (1982) also described a more structured approach to hydrological generalisation that would be superior to the 'spaghetti like disorder' present in many cartographic databases. Although a cartographer might have a good aesthetic sense, researchers in other disciplines can offer better knowledge about what should be preserved at reduced scales. In the near future, there will be enormous scope for specialists to use automated generalisation in order to tailor maps to their own individual specifications. This could revolutionise the production of specialised digital maps.

This raises the issue as to whether manually generalised maps should continue to be used as a quality criterion for automated generalisation. The quality of computer-produced maps is often determined by comparing the results with manually produced maps (Müller *et al.*, 1995). While manual cartographers carried out much of the early work on generalisation, the advent of computer-based mapping techniques has called into question the need to adhere to the pre-existing conventions (Rhind, 1973). The need for generalisation has changed somewhat as many map products made using the computer have become single-purpose rather than multi-purpose. While authors such as Jenks (1979) have suggested that the concepts and principles of manual cartography should not be neglected when implementing automated procedures, other authors (e.g. Rhind, 1973) argue that the principles and practice of generalisation of cartographic elements in an automated system should not be assumed to be the same as those in manual cartography.

However, as long as automated generalisation lags behind the manual generalisation of expert cartographers, then an intermediate system (such as the amplified intelligence approach described in section 1.3.3), which relies on heavy input from the user, is indispensable. In parallel to the development of better generalisation algorithms, it is important to devise mechanisms that can provide greater user-control over effects resulting from automated map manipulation processes. Such mechanisms should allow the user to control the generalisation effects of individual features, while maintaining the correct topological relationship between the different features. This is the driving force behind the development of new context-dependent generalisation algorithms within a feature-based environment, as discussed in chapter 1. This study has also pointed out that some features are more affected in particular ways than others (e.g. displacement rather than length change). Therefore generalisation effects depend not only on the feature type but also on the aspects of the feature with which the user is most concerned. This demonstrates the importance of the user controlling the effects of generalisation. A possible framework for such a system is proposed in the following section.

7.3 Control of generalisation effects within GIS

At present, GIS systems fail to offer sufficient support to users who wish to apply generalised data or to generalise existing data further. In particular, as

João *et al.* (1993) have pointed out, users often have to take important decisions relating to generalisation that affect the accuracy and 'quality' of the data, without any guidance as to the possible consequences of these decisions for the data, or access to any technology for identifying the resulting spatial data uncertainties. This is especially serious because, as Openshaw *et al.* (1991) have shown, errors contained in digital maps propagate during a series of GIS operations. Ultimately, the use of GIS as a decision-making tool will be reduced if the accuracy of the results cannot be controlled.

While manual generalisation effects might be reduced by the use of larger-scale maps, it is also important to devise mechanisms that can provide greater user-control over generalisation effects due to automated generalisation processes. This is especially important if a scale-independent GIS, as described in section 1.4, is ever developed, because in this eventuality data would always be generalised by automated methods. In a scale-independent GIS, data would be stored at the finest resolution possible, and generalisation algorithms would then be used to generalise the data whenever a smaller-scale application was required.

While the more ambitious scale-independent GIS remains beyond the capabilities of present-day computer systems, users of databases that store multiple representations (such as those in today's mapping agencies) should use automated generalisation methods as much as possible. Since automated methods appear to cause fewer generalisation effects than manual methods, as demonstrated in this study, this could be a way of reducing the effects of generalisation. Therefore, even with the multiple-representation type of spatial database, the development of methods to quantify and control generalisation effects is still a priority. However, a GIS user faces considerable difficulty in controlling generalisation effects caused by automated generalisation methods, since different algorithms cause variable effects on the data (McMaster, 1987) and, as shown in this study, the magnitude of these effects will vary for different features.

To control the extent of generalisation effects within a GIS, the development of a new type of automated generalisation system is proposed. This new automated system for generalisation would allow the user to control generalisation effects in a two-stage process. An *a priori* control would let the user set quality constraints on the generalisation process (i.e. by specifying the type and amount of generalisation effects acceptable), and an *a posteriori* check would give the user an opportunity to assess the quality of the end result through the analysis of maps and reports. According to this principle, the automation of generalisation within a GIS would be approached as an *optimisation process*. Generalisation facilities available within a GIS would allow the selection of algorithms that would minimise particular generalisation effects and therefore optimise the process from the point of view of the user.

Within this proposed system, and similar to the amplified intelligence approach described in section 1.3.3, the user would play a key role. However, unlike any existing generalisation system, a major concern would be in explicitly reducing generalisation effects. The users would be instrumental in the control of generalisation effects, as it would be their accuracy needs that would set the main constraints for the generalisation procedure. This sort of user intervention in automated generalisation, based on controlling the *extent* of generalisation effects in the first place, has been stressed previously by the author and by research colleagues (see João, 1991; João *et al.*, 1993).

The facilities available in commercial GIS for the user to control generalisation processes are very limited, despite the risk of producing inaccurate geographical data through the misuse of generalisation operations. The difficulty of automating generalisation poses the real danger that a completed generalisation system may perform merely *what is currently possible* rather than *what is needed* or *what is desirable*. After some time, the existence of such systems could set *de facto* standards that might blinker the user's perception of what is needed. It is therefore important to propose the existence of a new set of generalisation functions that are required from the user's perspective, independent of the viability of their implementation in the short term.

7.3.1   Control of effects before starting to generalise

The control of generalisation effects can start even before the generalisation process begins. New GIS functions would be useful that give advice to the user about whether generalisation is required or advisable. It is important to define the limits of what it is possible and sensible to attempt, and to educate the user about these limits: for example, situations in which it is inadvisable to combine a data set at a particular scale with data from a very different scale. The basic functions that need to be available to the GIS user before starting to generalise are displayed in Figure 7.1. The functional requirements are represented in the figure using a *user–concept diagram*. This is a type of data flow diagram in which automated processes are identified according to the user's view and the movement of data is traced through the processes (Martin, 1988). Such diagrams give special attention to the identification of anticipated system uses and how users will interact with the system.

The most elemental function in Figure 7.1 is number 1. This allows the map content to be displayed and queried, and data quality statistics and lineage (such as source scale, accuracy, date and source) to be made explicit. This function should be the basis for map selection, whether for display or analysis purposes, equivalent to cartographic and model generalisation. Although this function is already available in most GIS, it usually lacks the listing of data quality statistics which are integrated with map display. Coupled with function 1 would be function 2, which permits the user to query the need to generalise a particular map, or the benefit from so doing (based, for example, on the mea-

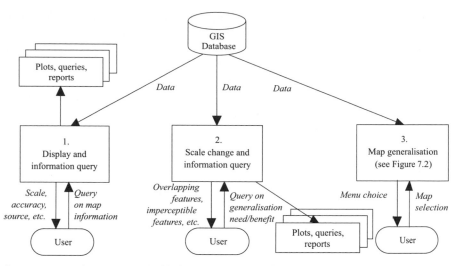

Figure 7.1 GIS functions needed before generalising (João, 1991).

sures – proposed by Shea and McMaster (1989) – that indicate when generalisation is needed).

On the basis of the change of scale required, the GIS could give the user a 'pre-generalisation report' that would list the proportion of features that would become imperceptible on scale reduction, the number and type of features that could overlap, the trouble areas of the map where features could become cluttered, and so on. Ideally, this information would be available even before the time-consuming process of generalisation had taken place. This would be possible if knowledge about past generalisation of similar maps, or information on established theoretical values (e.g. the minimum mapping unit for different output scales) was available. A more intelligent system could inform the user which type of generalisation would be required (e.g. line simplification) and for which type of features (e.g. coastline). The user could then obtain information on these types of generalisation and learn about the changes that could affect the data (e.g. vector displacement). On the basis of the information given by functions 1 and 2, the user could then decide to perform the generalisation itself (function 3, shown in more detail in Figure 7.2 below).

7.3.2 Control of effects during the generalisation process

The second step in the control of generalisation effects within GIS systems would be to minimise the extent of generalisation effects; for example, by avoiding unnecessary over-simplification of a line. This would be achieved by an automated generalisation system which would select generalisation processes on the basis of the minimisation of the generalisation effects as specified by the user. These specifications would ideally be based on the type of generalisation descriptors used in this book. Any such system would rely on the user to supply information about the purpose, the scale required, the quality tolerances, the relevant map information and the priority that should be taken into account during generalisation.

On the basis of this information, the proposed automated generalisation system would select – or help the user to select – the most appropriate generalisation algorithms, the best generalisation tolerances, and the recommended sequence in which to apply the different algorithms (if more than one type were selected). However, in the initial stages of the development of such a system, it is likely that the user would be required to make certain decisions, such as which algorithms would be used to generalise the data. Eventually, and similarly to the strategy proposed by Weibel (1991), as many as possible of these decisions would be carried out automatically by a knowledge-based system. The approach would be to develop programs incrementally, permitting the inclusion of further details on the effects of generalisation for different generalisation operators and for different classes of features as knowledge is accrued.

In Figure 7.2 is shown the user–concept diagram for the different functions for generalisation that a future GIS would ideally have: the user interaction with each of these functions is identified. When users choose to generalise a map, the initial query concerns the map's purpose (function 3.1). A list of default purposes could be available and any new purposes could be added as necessary. In a GIS environment, and as discussed in section 1.3, the two main types of purpose are for display or analysis. Beyond this choice, the user could also be asked to describe for which specific purpose the map is going to be used (e.g. hydrology). This is important in helping to determine the appropriate scale, the priority and the importance of features (e.g. rivers in hydrological maps), and which measures of generalisation effects will be most appropriate.

On the basis of the purpose, the system would ask the user to choose the new scale required (function 3.2). Choosing an appropriate scale for analysis can be more complex than choosing an appropriate scale for display, as it is dependent both on the uncertainty about the accuracy of individual observations (Müller, 1989) and at which scales particular spatial processes occur (Müller, 1991b). A warning mechanism would advise if the scale for the specified purpose was not suitable (function 3.3). Next, the user could be asked to input map information that was relevant for the generalisation (function 3.4). The classification parameters should be included in GIS attribute tables. A classification of entity types could include a description of the geometric form and measurement level of the geographical features (which depend on scale); a classification of the geographical features according to their integration in a

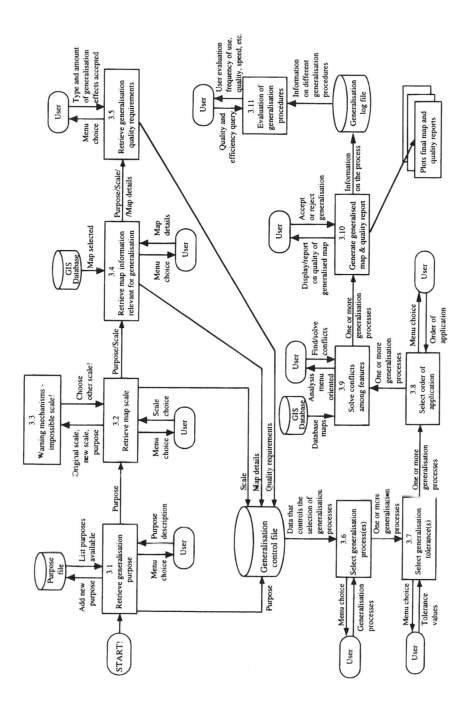

Figure 7.2 Functions for generalisation that GIS should possess (João, 1991).

geographical region (i.e. measures of density, spacing and interaction); and a classification of how each geographical feature relates to other types of geographical features – for example, rivers should remain perpendicular to contours (João et al., 1990).

Finally, and most importantly, the user would input the quality requirements for the generalisation; in other words, the maximum generalisation effect that the user accepts per entity type (function 3.5). The quality requirements in the case of model generalisation could include details such as the maximum acceptable vector and area displacement, and the maximum acceptable change of feature length. In the case of cartographic generalisation, the user would want to try to maximise the recognisability or natural look of the geographical features. In this case, either some sort of 'shape accuracy' measure would need to be developed, or alternatively the user would simply specify a 'natural look' and natural type algorithms would be applied (see, for example, Li and Openshaw, 1992; Müller and Wang, 1992). Within this function the user would also identify an hierarchy of feature importance, so that the conflicts between features (arising from a concentration of features at smaller scales) could be resolved. For consistency, these priorities could also guide the sequence of application of the algorithms to each individual geographical feature.

Together, the information on purpose, scale, map details and quality requirements constitute the generalisation control file (see Figure 7.2). The generalisation control file is the kernel of the system. This file would control the selection of generalisation algorithms to be applied to the data (function 3.6). The system should be designed in such a way that the algorithms selected should conform to the user's quality requirements; that is, they would minimise certain generalisation effects. The choice of generalisation algorithms would be followed by the selection of the generalisation tolerances for the respective algorithms (function 3.7). The order in which these algorithms are applied would also be taken into account (function 3.8). For example, if both a simplification and smoothing algorithm were selected, the simplification algorithm should be applied first (as recommended by McMaster, 1987).

At this point, any possible conflicts between different types of geographical features would be reported (function 3.9). For example, if a forest map is being generalised, the user needs to confirm that the amalgamation of forest areas is not going to overlap any rivers running between the amalgamated areas. It is very important that geographical features are not generalised in isolation or in the abstract. In the case of GIS in which different geographical features are stored in separate layers (e.g. ARC/INFO map coverages), each layer is generalised independently of all other layers. This poses complex problems, and is one of the reasons why the use of feature-based systems has been recommended for generalisation (see section 1.4). Nevertheless, mechanisms need to be found for layer-based systems that take into account the geographical relationship between features at the time of generalisation.

The classification of entity types in function 3.4 could help to solve this problem, but it is important that some sort of conflict resolution exists (function 3.9). The user can be asked to solve specific conflicts interactively and assign priorities to certain features. For example, if a map for display purposes is needed, and a railway, a road, a river and a power line all run through the same narrow valley, which feature should be displaced the least? When the generalised map is generated (function 3.10), this should be coupled with a quality report that informs the user about the changes created in the data. Only after this quality report has been produced can the user make an informed decision about whether to accept or reject the generalisation carried out.

7.3.3   Control of effects after generalisation is completed

After generalisation has been carried out, the user will normally have available a graphical representation of the simplified data set (at least in the case of cartographic generalisation). However, other reporting facilities are also needed to verify whether the generalisation requirements have been met. The system should report back on changes resulting from generalisation, and should indicate how closely it has been able to meet the user's specifications by giving specific values for length change, point loss, and so on. In addition to this, there may be occasions when a user wishes to know, for a given set of measurements (examples being length change or vector displacement), which changes have taken place to the data, so that there is a record of the extent to which a particular file has been altered. If some excessive effects of generalisation are unavoidable, damage limitation is still possible if the extent of these effects is made known to those who use the data in the future. This could help to minimise errors, described by Beard (1989), due to the misuse of spatial data and operations by GIS users. Without information on data quality it is all too easy for data to be used for a purpose for which they were not originally intended (Blakemore, 1985).

Even after generalisation has been performed, the user can still reduce the impact of generalisation effects by developing ways of taking into account the *quantified* generalisation effects in GIS map manipulations. At the very least, the data could be tagged in terms of the transformations suffered in the generalisation process. Associated descriptors of likely generalisation effects (e.g. changes in line length), when converted into confidence levels, could be used to classify the accuracy of results and avoid misuse of information. Further research on developing error and uncertainty estimation models, and embedding these in error handling procedures related to GIS is therefore fundamental (Goodchild and Gopal, 1989; Brunsdon and Openshaw, 1993).

If the generalisation carried out is rejected (because it caused too many unwanted changes in the data), the user would be asked to input different quality requirements or, alternatively, the user could choose to re-initialise the whole process (see Figure 7.2). In either case, the steps to perform the generalisation process could be stored in a *generalisation log file*. By storing information describing the generalisation processes performed for each specific situation, it will be possible to repeat these processes for more than one map (e.g. for the production of an entire map series) or avoid potential pitfalls. The generalisation log file could also store information on the efficiency and quality of each generalisation process; for example, the speed of the procedure, the number of iterations until the user was satisfied with the result, and so on. Based on this information, an additional function is available for evaluating the generalisation procedures within their own terms of operation (function 3.11). This function allows the user to compare different generalisation procedures performed at different times or for different types of data.

The improvement of the automation of the generalisation process should be closely coupled with the effective control of generalisation effects. João *et al.* (1993) proposed an automated generalisation system that increased the level of automation in generalisation while simultaneously optimising the quality of the final product. A general solution to minimise generalisation effects would then be for a single topographic base to be used whenever possible, for data to be retained at as detailed a scale as possible, and for automatic generalisation to be used only when needed. The extent of cartographic generalisation effects would be minimised by the use of large-scale maps – larger than necessary for any one application. The control of the effects resulting from automated generalisation processes would be achieved by the use of a generalisation system with the capabilities just described.

7.4   The future

This study has taken one particular approach to understanding the extent of generalisation effects. It has quantified the problem for two families of maps for the first time, and simultaneously confirmed the necessity of controlling generalisation effects within a GIS. However, there are many more approaches to the problem of generalisation, and the aim of future research must be to corroborate some of the findings of this research. While cartographically generalised maps continue to be digitised and used in GIS analyses, the study of the potential errors introduced by using these generalised maps should be a priority.

While this study has looked at two localised and similar areas in considerable detail, it would have been possible to look at a wider range of study areas in less detail. For example, it will be important to quantify the generalisation effects of very different topographies. Mountainous areas could be compared with coastal plains. It seems likely that the more mountainous the region is, the more generalisation might occur, but this needs to be confirmed quantitatively using the descriptors and methodology that have been developed for this study. Similarly, it would be possible to extend the effects of generalisation to maps of a smaller scale than the 1 : 625 000 and larger than the 1 : 50 000 scale maps used in this research.

Another approach would be to look at the generalisation inherent in non-topographic maps; for example, hydrological maps. General topographic maps are multipurpose maps that can be used for a wide range of applications. This makes it difficult to pin-point the importance of generalisation effects because often it will depend on the application of the map. However, if specific applications (e.g. hydrology) are analysed, then the generalisation effects could be analysed in relation to the particular aims of hydrological studies. On hydrological maps it would be preferable for water courses to retain greater positional accuracy than roads and, correspondingly, for roads to be generalised more.

In general, the full impact of generalisation effects is apparent when GIS map manipulations are carried out, and further emphasis should be placed on this. GIS are invaluable for carrying out complicated geographical analyses, but it is vital that users are aware of the limitations of using generalised data sets. Geographical analyses study phenomena that manifest themselves in space, implying a focus on location, area, distance and interaction (Anselin, 1989). Generalisation affects all these variables to a lesser or greater extent, with potentially dramatic effects on the results of any GIS map manipulation. While this study used theoretical GIS map manipulations in order to keep control of the variables, real case studies could also be examined. The author is presently starting a project to carry this out. Various real case environmental impact assessments (EIA) are being repeated using different source scale maps. Performing the same analysis across different scales will help to highlight the effects of generalisation on the outcome of EIA.

There is a growing awareness among GIS users about the distinction between model and cartographic generalisation (see section 2.2). In a GIS context, it is often just as important to take account of the *statistical* consequences of generalisation as it is to take account of the *graphical* consequences – and the generalisation effects involved will not necessarily be the same for each case (João *et al.*, 1993). Because cartographic generalisation has been around for a much longer period of time (and because most spatial databases are digitised from existing maps), considerable research on it has been carried out, but eventually much more work will be needed on the effects of model generalisation and the relationship between the two. A possible research question could be to investigate whether optimal model generalisation is incompatible with optimal cartographic generalisation (i.e. whether they are mutually exclusive).

The type and extent of generalisation effects are intimately related to the purpose of the generalisation process (i.e. cartographic versus model). Positional accuracy with regard to model generalisation is different from what can be called 'shape accuracy', which is of relevance to cartographic generalisation. Positional accuracy relates to how accurately the position of a geographical feature is recorded relative to its real-world position, while shape accuracy relates to how a shape of a feature is represented compared with its real shape (Smith and Rhind, 1993). In some instances there may be a conflict between the two aims: sometimes it may be necessary to sacrifice high positional accuracy in order to increase the recognisability of a geographical object to a human observer. Conflict may also occur between absolute and relative accuracy, when the absolute accuracy is reduced in order to maintain the correct topology. It is therefore important to distinguish between cartographic generalisation and generalisation for modelling purposes, particularly when referring to data quality. More research is also necessary on the effectiveness of measures of generalisation effects in controlling the quality of both model and cartographic generalisation.

A related distinction is between automated and manual generalisation. While they were compared in this study, the only automated algorithm used was the Douglas–Peucker. Any future study comparing manual with automatic generalisation must look at other types of algorithms. Future research should also involve the development of more sophisticated algorithms for model and cartographic generalisation. It is also important to perfect and develop more efficient generalisation algorithms that minimise generalisation effects. A GIS that bases the selection of generalisation algorithms on quality requirements should be complemented by a wide range of available algorithms. If, for example, a user requires the simplification of a line with minimum vector displacement, the GIS should possess an algorithm that could optimise this. The distinct effects of generalisation for different algorithms therefore need to be fully identified and measured, building on work such as that of McMaster (1987) and Müller (1987).

Coupled with the above is the need for better generalisation functions within GIS. Few GIS available on the market (with the exception of the Map Generalizer and LAMPS2 – see section 1.3.3) have mechanisms to cope with generalisation-related problems except in the most trivial way, such as through 'weeding' of points within lines. In addition, none routinely produces a statistical summary of the changes to the overall characteristics of spatial data after such a transformation has been applied. An ideal generalisation procedure within a GIS would select the best available generalisation process based on purpose, scale, data types and quality requirements. Also, it would keep track of the changes upon the data by the generalisation procedures. The aim, ultimately, is to rely as much as possible on a knowledge-based system to carry out this process for the user. However, at the moment, knowledge-based generalisation is severely limited and so any successful generalisation within a GIS will need to depend heavily on the input of the user.

Recent research into generalisation (including this study) has led to an increasing awareness of the need to re-evaluate what constitutes an effective map at each scale. Eventually, it might even be necessary to re-generalise existing maps. In the past, the ideal objective of automated generalisation has been to replicate the best maps produced manually, but this assumes that manually produced maps are the optimal solution. The highest standard that automated generalisation should aspire to is successfully to control the quality of the generalised cartographic products. This is not just quality control similar to that of the manual cartographer's check, but *quantification* and *storage* of the quality values that can then be carried forward and taken into account in subsequent GIS map manipulations. The aim of automated generalisation should be to maximise accuracy, rather than merely to mimic manual methods. It might now be necessary for mapping agencies progressively to re-draw their small-scale maps using more automated generalisation and/or according to the specifications of the users (rather than according to the specifications laid down by the data supplier). François Salgé, for example, has reported that the French Survey (the Institut Géographique National – IGN) is at present creating new maps for GIS use (using an 'object-approach') rather than digitising existing analogue maps (which were done using a 'drawing-approach'): 'The object-based databases (of the IGN) are characterised by the representation of the landscape as digital information rather than a digital representation of cartographic drawings' (Salgé, 1993, p. 2).

Although this study has some relevance to conventional cartographic techniques, it is ultimately of greater relevance to the GIS community, where the ability to quantify generalisation could be made possible by automation and where the consequences of uncontrolled generalisation are more serious. The proposal in this book to control the effects of generalisation has focused on the development of a user-controlled program in which the user is able to give instructions concerning the purpose, output scale, quality requirements and other map information relevant for map generalisation (see section 7.3). Central to this philosophy is the initial design of a generalisation machine that can cope with all the known generalisation processes (in combination wherever appropriate), but in which the modules are progressively implemented, as resources and knowledge permit.

The potential contribution of genetic algorithms or neural networks for the implementation of such a system could be investigated, building on research such as that of Keller (1995) and Baeijs et al. (1996). Although these sort of procedures are computationally intensive, current trends in the improvement of hardware speed suggest that this might not be a limitation in the future (João et al., 1993). Given the necessity for 'intelligent' systems to explain how they have arrived at a given conclusion, it would seem possible to extend the notion of explanation to include the ability to monitor the effects of the actions taken on the data. This would also allow the introduction of facilities designed to minimise the occurrence of undesirable effects through the specification by the user of accuracy constraints as part of the goals to which the system would have to adhere.

The approach proposed in this book maintains the user as the main agent in specifying objectives and setting constraints. The focus of automation then becomes one of finding the best ways in which the user can carry this out and then of letting the system decide the optimal way of achieving the user's requirements. This latter aspect means that there must be at least some rudimentary decision-making mechanism built into generalisation systems to enable choices to be made between alternative actions. Even more important is a form of *learning mechanism*, so that the knowledge a system has of different situations or strategies is progressively increased. Such a system need only be provided with the structure of the knowledge that it requires, rather then having to be pre-programmed with every rule for every possible situation. This type of mechanism, as reported by Weibel (1991) and Herbert *et al.* (1992), may represent the only strategy for overcoming the potentially vast number of rules which would be needed to initiate an 'intelligent' generalisation system that was capable of coping with a variety of tasks and differing data sets. Instead, the rules would be added as required. Knowledge-based systems are an alternative basis for progressing further with automated generalisation because they involve the transfer of human expertise, or other knowledge about the subject, to an automated program.

As a consequence of the above, a further research avenue would be improved user interfaces. Better control of the quality of the generalisation process may be possible if the user is able to customise the user interface. In this way, users could emphasise the control of particular quality parameters (e.g. preservation of angularity) that are most important for their applications. Error messages and warning indicators (such as the 'metaphors for conflict' proposed by Mackaness and Beard, 1990) could also be developed according to the user preferences. The customisation of such a generalisation control system could be possible through the interface facilities provided by a system such as UGIX (Rhind *et al.*, 1989; Raper and Bundock, 1991). Also, it is important that users are given an opportunity to define their own way of measuring generalisation effects. This would allow the user to vary the way in which the existing measures are usually applied, or even permit them to develop their own new measurements.

Independently of the development of sophisticated intelligent systems that would control generalisation effects, there is scope for improving the information given to users about the type and degree of generalisation contained in spatial data sets available at present. The information given by the Australian Survey in terms of the expected displacements for different feature types according to the number of features under conflict (Australian Survey, 1992) is already a step forward (see section 2.2.3). In addition, all digital data sets should be accompanied by a statement on data quality statistics that would indicate to the user the positional accuracy of the different features in different parts of the map. 'Grey boxes', for example, could appear around heavily generalised areas, to warn the user about potentially sizeable generalisation effects and possibly motivate the user to obtain larger-scale maps. The concept of 'fitness for purpose' can only be determined by the users themselves and not by the mapping agency producing the data. In order for the user to assess this fitness for purpose, specific quantitative measures of accuracy would be necessary rather than more vague qualitative statements.

In conclusion, this study found that generalisation effects are important both in terms of changes to individual features (a few of these effects were counter-intuitive) and also to the results of GIS map manipulations. It must be an important caveat to all users of GIS that they should be aware of the problems of generalisation when they carry out any spatial analysis procedure. Ultimately, it is hoped that the generalisation effects will be minimised by developing a system such as the one proposed in this final chapter. This system should differentiate between cartographic and model generalisation. It is inevitable that the generalisation effects will be larger in the former case, but in any event the size of the effects should be closely monitored and controlled by the user. Further research will include widening the scope of this study, as well as the concrete development of the proposed system to control generalisation effects within a GIS.

Generalisation effects embedded in the British 1 : 50 000 scale map

The 1 : 50 000 scale maps for both study areas were used as a baseline from which the generalisation effects of the smaller-scale maps were determined. For the Portuguese study area, the 1 : 50 000 map is the basic scale map. However, since the 1 : 50 000 map for the British study area is derived from the 1 : 10 000 scale map, it was possible to investigate the extent to which the 1 : 50 000 map had been generalised in relation to this scale. This appendix shows the results of this investigation. There are two major sections in this appendix: changes in line length and lateral displacement of features.

The location of the 1 : 10 000 scale map sheet used (TR 15 NW) in relation to the 1 : 50 000 scale map is shown in Figure A.1. The figure only shows line features (i.e. railways, river, roads and boundaries) of the 1 : 50 000 scale map which are common across the four scales.

The roads, railways and river of the 1 : 10 000 scale map used in the comparison with the 1 : 50 000 scale map can be seen in Figure A.2. In order to make a valid comparison of the maps, the features of the 1 : 50 000 scale map had to be cut, using the area occupied by the 1 : 10 000 scale sheet as a template. In Figure A.2 it can be seen that there are two railway lines: the one passing south of Canterbury goes to Dover and the other passing north of Canterbury goes to Ramsgate. There are five roads within this map sheet that are common across all the scales (1 : 10 000, 1 : 50 000, 1 : 250 000 and 1 : 625 000). All of the roads were represented on the 1 : 10 000 scale map as a double line, and therefore the centreline was digitised. For the sake of this study, the group of roads that encircle Canterbury's city centre is called the 'circular road'. This circular road is in fact composed of a series of independent roads (e.g. Rheims Way, Pound Lane, Broad Street, etc.). Note that the right border of the 1 : 10 000 map sheet partially cuts this circular road (see Figure A.1). The A2050 road is a slip road that links the A2 with this circular road.

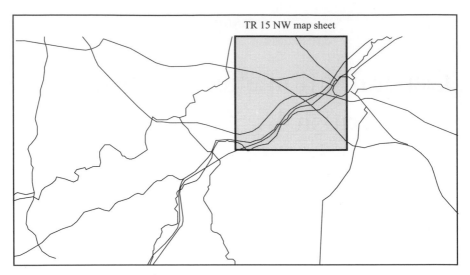

Figure A.1   The location of the 1 : 10 000 scale map sheet on the 1 : 50 000 OS map.
© Crown copyright.

Finally, there is the Great Stour river, which crosses the study area. The bottom edge of the 1 : 10 000 scale map cuts the river into two pieces. Therefore, although it is the same river, in the tables and figures that follow, the two pieces are referred to as two independent rivers.

Measurements of line length

There is some variation in the change of length of features between the two scales – see Table A.1 and Figure A.3. For most of the features there is very little change in length (not more than 0.7 per cent), but three of the features have a particularly large change in length. A small section of the Great Stour (in the bottom left corner of Figure A.2) decreases in length the most, by 8.5 per cent. However, this change of length is not so much due to the change in the detail of the feature but results from the displacement of this particular section of the river in a southwesterly direction (this displacement will be analysed further in the next subsection). As the river is displaced, it is shifted closer to the corner, and it is shortened at both ends where the study area is cut. The effect is exacerbated by its very short original length. The A2050 is longer at 1 : 50 000 scale. This was mainly due to difficulties in digitising the overpass link with the dual carriageway (the A2). The definition of the centre of the road was complicated by the slip road. Finally, the circular road has a small step where it joins the A290 (at 1 : 10 000 scale), which disappears at the

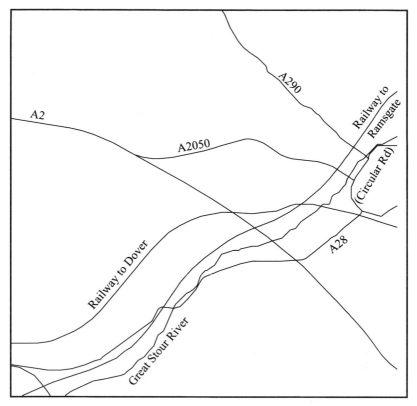

Figure A.2   The river, railways and roads (common to all the scales) represented on the 1 : 10 000 scale map TR 15 NW. © Crown copyright.

Table A.1   The length change of each feature at the scales of 1 : 10 000 and 1 : 50 000.

Features	Length on the 1 : 10 000 scale map (m)	Length on the 1 : 50 000 scale map (m)	Absolute difference (m)	Length change (%)
Rd ac 9 (A2)	6097.8	6098.9	1.1	0.0
Rd ac 13 (A28)	5283.5	5307.4	23.9	+0.5
Rd ac 17 (A2050)	3013.2	3150.4	137.2	+6.0
Rd ac 21 (A290)	2774.4	2770.4	4.0	−0.1
Rd ac 23 (circular road)	1794.3	1736.8	57.5	−3.2
Rail ac 2 (to Ramsgate)	6425.8	6397.5	28.4	−0.4
Rail ac 3 (to Dover)	5691.9	5701.2	9.2	+0.2
Rv ac 24 (Great Stour)	5783.0	5745.3	37.7	−0.7
Rv ac 25 (Great Stour)	668.5	611.4	57.1	−8.5

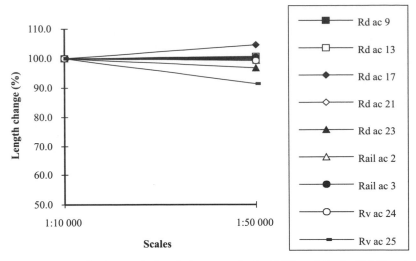

Figure A.3   Inter-scale changes in length, between the 1 : 10 000 and the 1 : 50 000 scale maps, of all of the individual features.

smaller scale. This accounts for the decrease in length. While this last case is a clear-cut case of generalisation (where detail is lost at the smaller scale), the other two examples have more to do with definition problems than with generalisation.

When averaging the differences by feature type (see Table A.2 and Figure A.4), the inter-scale changes in length are less pronounced. On average, roads and railways suffer the least changes in length. The rivers suffer a 1.5 per cent shortening, mainly due to the shorter corner section of the river, as described above. This would seem to indicate that the 1 : 50 000 map is a fairly faithful representation of the 1 : 10 000 map in terms of length measurements.

Lateral displacement of features

The displacements between the same features of the 1 : 10 000 and 1 : 50 000 scale maps are shown in Figure A.5. This figure gives a qualitative impression of the extent of feature displacement. It can be seen that features were minimally displaced, except for small sections of lines that lost detail locally and the river in the bottom left-hand corner, which was displaced along its entire length. In quantitative terms, in Table A.3 are given the different displacement measures for the individual features. The total area displacement per length gives the sum of all the displacement polygons per unit length (of the 1 : 10 000 scale map) between the two scales. As with the change in length, the three features displaced the most (in terms of areal displacement) are the A2050 slip

Table A.2 The length change by feature class of the 1 : 10000 and the 1 : 50000 scale maps.

Feature class (no. of features inside brackets)	Total length on the 1 : 10000 scale map (m)	Total length on the 1 : 50000 scale map (m)	Total absolute difference (m)	Total length change (%)
Roads (5)	18 962	19 064	105	+0.5
Railways (2)	12 118	12 099	19	−0.2
Rivers (2)	6451	6357	94	−1.5

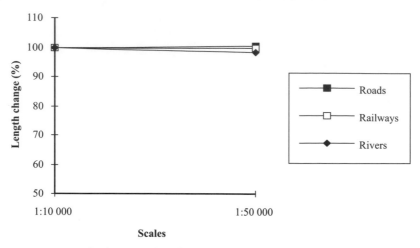

Figure A.4 Inter-scale changes in length of the roads, rivers and railways, aggregated per feature class, between the 1 : 10 000 and the 1 : 50 000 scale OS maps.

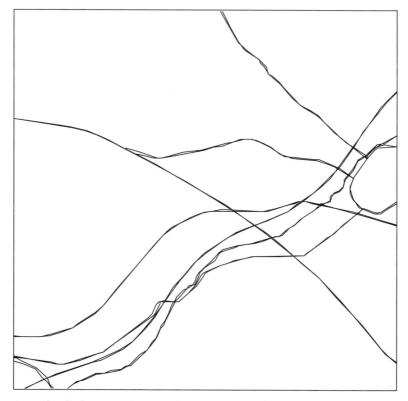

Figure A.5 The displacement between the same rivers, railways and roads represented on the 1 : 10 000 and the 1 : 50 000 scale maps.

Table A.3 The total areal displacement and number of polygons per length, and the maximum vector displacement between the 1 : 10000 and the 1 : 50000 scale maps for each feature.

Features	Total areal displacement per length (m²/m)	Number of polygons per length (no./km)	Mean size of each polygon (m²)	Maximum vector displacement (m)
Rd ac 9 (A2)	1.5	4.4	338.4	4.2
Rd ac 13 (A28)	7.1	3.4	2078.5	27.6
Rd ac 17 (A2050)	13.5	1.0	13 585.6	25.2
Rd ac 21 (A290)	6.9	4.0	1750.8	18.2
Rd ac 23 (circular road)	11.2	3.9	2861.4	29.8
Rail ac 2 (to Ramsgate)	5.9	2.0	2936.7	17.0
Rail ac 3 (to Dover)	4.6	2.8	1627.5	14.1
Rv ac 24 (Great Stour)	8.1	6.0	1333.7	27.5
Rv ac 25 (Great Stour)	28.3	1.5	18 894.9	34.6

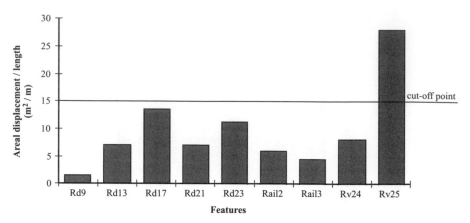

Figure A.6   The areal displacement per length of each feature.

road, the small section of the Great Stour and the circular road (see Figure A.6).

It is shown in Figure A.6 that, in mean terms, the only feature to have a displacement larger than 15 m was the small section of the river that was shifted sideways along its entire length. As described in section 4.1.2, 0.3 mm is taken as the cut-off point for the displacement of the 1 : 50 000 scale map corresponding to 15 m on the ground at this scale. Below that value, it is more difficult to determine whether the error is due to generalisation or to other sources. Displacement greater than 15 m has a large likelihood of being caused by generalisation. When features are aggregated, their displacement values all fall below the 15 m cut-off point (see Figure A.7).

The changes of length and displacement are linked. It was the displacement of the small piece of the Great Stour that caused its large change in length, by forcing both ends of the river to be cut shorter at the edge of the map sheet. In the case of the slip road, the displacement of the link with the A2 (due to definition-type difficulties during digitising) led also to both the displacement value and the change in length. In the case of the circular road, it was the elimination of the small step in the road at the smaller scale that mainly contributed to the observed displacement and change in length.

The number of polygons per unit length is greatest for the long section of the Great Stour (see Table A.3). Also, the mean size of each polygon is one of the smallest (only the A2 road has a smaller mean polygon size). This is because the river has more 'wigglyness' and, as found by Goodchild (1980b), the more detail a line has, the more it will generate sliver polygons when two different generalised versions of it are overlaid. In contrast, the small section of the Great Stour generates only one polygon, as it is completely displaced. As a

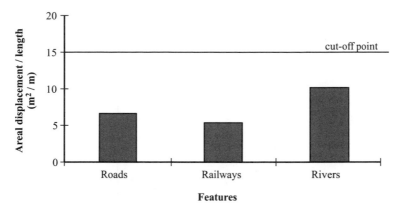

Figure A.7 The total areal displacement for each feature class divided by the total length of the features.

consequence, the mean size of the polygons generated is the largest for all of the features. Another measure of displacement is maximum vector displacement, which pin-points 'trouble spots'. As the name indicates, this measure shows the maximum amount of displacement for each feature (for example, where the circular road lost the small step mentioned earlier). As can be observed from Figure A.8, most of the features (seven in total) have particular sections that are displaced by more than 0.3 mm (i.e. 15 m on the 1 : 50 000 scale map).

The maximum vector displacement measures the maximum displacement in sections of the features, but does not necessarily measure exactly the same point across the scales. Only by using well-defined points can the displacement across scales of exactly the same point be measured. In order to analyse the

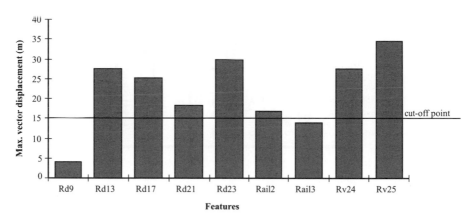

Figure A.8 The maximum vector displacement for each feature.

Table A.4 The displacement of intersection points between features.

Features	Rd ac 9 (A2)	Rd ac 13 (A28)	Rail ac 2 (to Ramsgate)	Rail ac 3 (to Dover)	Rv ac 24 (Great Stour)
Rd ac 9 (A2)					
Rd as 13 (A28)	10.8 m (0.22 mm)				
Rail ac 2 (to Ramsgate)	3.8 m (0.08 mm)	5.1 m (0.10 mm)			
Rail ac 3 (to Dover)	7.4 m (0.15 mm)	11.2 m (0.22 mm)	7.7 m (0.15 mm)		
Rv ac 24 (Great Stour)	3.5 m (0.07 mm)	17.0 m (0.34 mm)	30.7 m (0.61 mm)	15.5 m (0.31 mm)	

displacement of well-defined points, the displacement of intersection points was measured. In Table A.4 is shown the displacement of the intersection points between some of the features: the two railway lines, the two main roads (the A2 and the A28) and the large section of the river.

The displacement of all of the road intersections, and of road intersections with railways, generated displacements that were all below the cut-off point. Intersections with the Great Stour generated both the smallest displacement (in the case of the intersection point with the dual carriageway A2) and the largest value (in the case of the intersection with the railway line to Ramsgate). Only the intersections of the river with the two railway lines and with the A28 generated displacement values above the cut-off point of 0.3 mm.

In summary, the differences in line length and displacement were varied. Some of the individual features showed a maximum displacement greater than the cut-off point. However, when aggregated together (in terms of feature length and displacement), the differences due to generalisation were found to be small. This supports the use of the 1 : 50 000 scale map as a baseline.

Raw data for each feature per scale for the two study areas

In this appendix, nine tables containing raw data are presented. The Portuguese study area is detailed in Tables B.1–B.4, and the British study area in Tables B.5–B.9.

Table B.1 Length and sinuosity for the Portuguese study area (scales 1:50 000, 1:200 000 and 1:500 000).

Scales	Roads				Railway	Rivers		
	rd ac1 (245)	rd ac35	rd ac36	rd ac41 (243)	rail ac2	riv ac32 Rib. Grande	riv ac175 Rib. de Lupe	Boundary bound ac7
Length (m)								
1:50 000	10 462.65	2022.52	11 064.85	10 275.76	14 154.41	21 879.22	1952.10	16 237.25
1:200 000	10 071.21	1738.16	11 050.22	10 186.49	13 846.49	20 582.10	1791.58	15 125.23
1:500 000	9705.66	1888.52	10 908.45	10 065.70	13 518.43	19 369.15	2028.37	12 493.80
Number of points (=n)								
1:50 000	123	28	98	115	222	592	48	199
1:200 000	57	11	66	57	67	168	26	91
1:500 000	16	8	24	33	39	176	19	52
Average segment length (=length/(n − 1)) (m)								
1:50 000	86	75	114	90	64	37	42	82
1:200 000	180	174	170	182	210	123	72	168
1:500 000	647	270	474	315	356	111	113	245
Cumulative angularity (degrees)								
1:50 000	850.45	104.04	616.17	617.14	1016.29	3383.08	807.47	2779.70
1:200 000	436.48	71.30	402.07	499.87	688.22	1861.89	244.66	1658.88
1:500 000	66.54	50.94	154.70	338.12	565.35	1766.85	144.64	714.25
Angularity per unit length (degrees/km)								
1:50 000	81.28	51.44	55.69	60.06	71.80	154.63	413.64	171.19
1:200 000	43.34	41.02	36.39	49.07	49.70	90.46	136.56	109.68
1:500 000	6.86	26.98	14.18	33.59	41.82	91.22	71.31	57.17
Average angle (=cumulative angularity/(n − 2) (degrees))								
1:50 000	7.03	4.00	6.42	5.46	4.62	5.73	17.55	14.11
1:200 000	7.94	7.92	6.28	9.09	10.59	11.22	10.19	18.64
1:500 000	4.75	8.49	7.03	10.91	15.28	10.15	8.51	14.29

Table B.2 Displacement measures for the Portuguese study area (scales 1 : 50 000, 1 : 200 000 and 1 : 500 000).

	Roads				Rivers			
Scales	rd ac1 (245)	rd ac35	rd ac36	rd ac41 (243)	Railway rail ac2	riv ac32 Rib. Grande	riv ac175 Rib. de Lupe	Boundary bound ac7
Total areal displacement (m²)								
50k–200k	346 990.4	97 567.2	471 747.8	541 174.7	719 277.1	790 859.2	123 673.8	889 250.8
50k–500k	1 340 705.0	329 061.9	5 668 189.0	1 163 827.7	3 006 198.5	3 618 666.3	233 261.3	3 217 138.0
Number of polygons								
50k–200k	12	3	9	5	11	37	1	22
50k–500k	7	2	7	9	4	11	1	12
Largest polygon (m²)								
50k–200k	122 164.5	60 840.9	213 366.9	251 686.5	413 110.6	123 222.0	123 673.8	196 489.2
50k–500k	449 788.5	217 269.5	5 562 023.5	345 651.5	2 956 923.7	879 054.5	233 261.3	1 259 666.8
Smallest polygon (m²)								
50k–200k	558.1	1533.4	20.7	6209.0	2809.5	0.1	123 673.8	62.1
50k–500k	1556.7	111 792.4	21.3	387.3	1185.3	91.5	233 261.3	13 718.9
Average polygon size (=total areal displacement/no. polygons (m²))								
50k–200k	28 915.9	32 522.4	52 416.4	108 234.9	65 388.8	21 374.6	123 673.8	40 420.5
50k–500k	191 529.3	164 531.0	809 741.3	129 314.2	751 549.6	328 969.7	233 261.3	268 094.8
Maximum vector displacement (m)								
50k–200k	260	105	142	139	131	177	144	155
50k–500k	404	289	983	218	538	229	516	1110
Total areal displacement per unit length (m²/m)								
50k–200k	33.2	48.2	42.6	52.7	50.8	36.1	63.4	54.8
50k–500k	128.1	162.7	512.3	113.3	212.4	165.4	119.5	198.1
Number of polygons per unit length (polygons/km)								
50k–200k	1.1	1.5	0.8	0.5	0.8	1.7	0.5	1.4
50k–500k	0.7	1.0	0.6	0.9	0.3	0.5	0.5	0.7

Table B.3 Comparing manual and automated generalisation for the Portuguese study area (length and sinuosity for the scales 1 : 50 000, 1 : 200 000 and 1 : 500 000).

	rd ac1 (245)	rail ac2	river ac32 (Rib. Grande)	bound ac7
Douglas–Peucker algorithm tolerance (DP) (m)				
1 : 50 000 (crucial)[a]	0.01	0.1	0.4	0.1
1 : 200 000 (crucial)[a]	0.8	0.9	0.3	1
1 : 500 000 (crucial)[a]	2	3	0.75	2
1 : 200 000 automatically generalised[b]	3.65	8.6	6.5	9.3
1 : 500 000 automatically generalised[b]	50	26.8	6.8	26
Number of points (=n)				
1 : 50 000 (crucial)[a]	121	219	592	196
1 : 200 000 (crucial)[a]	52	60	164	88
1 : 500 000 (crucial)[a]	15	35	161	49
1 : 200 000 automatically generalised[b]	52	60	164	88
1 : 500 000 automatically generalised[b]	15	35	161	49
Length (m)				
1 : 50 000 (crucial)[a]	10 462.646	14 154.406	21 879.221	16 237.253
1 : 200 000 (crucial)[a]	10 071.202	13 846.490	20 582.100	15 125.226
1 : 500 000 (crucial)[a]	9705.659	13 518.428	19 369.150	12 493.771
1 : 200 000 automatically generalised[b]	10 457.206	14 131.696	21 828.943	16 210.318
1 : 500 000 automatically generalised[b]	10 383.797	14 090.204	21 827.244	16 112.699
Average segment length (m)				
1 : 50 000 (crucial)[a]	87.189	64.928	37.021	83.268
1 : 200 000 (crucial)[a]	197.475	234.686	126.271	173.853
1 : 500 000 (crucial)[a]	693.261	397.601	121.057	260.287
1 : 200 000 automatically generalised[b]	205.043	239.520	133.920	186.325
1 : 500 00 automatically generalised[b]	741.700	414.418	136.420	335.681
Cumulative angularity (degrees)				
1 : 50 000 (crucial)[a]	849.514	1015.914	3383.076	2779.411
1 : 200 000 (crucial)[a]	436.186	687.509	1861.854	1658.692
1 : 500 000 (crucial)[a]	66.362	565.340	1764.765	713.954
1 : 200 000 automatically generalised[b]	711.896	842.848	2738.952	2399.630
1 : 500 000 automatically generalised[b]	535.441	793.160	2730.030	2125.741
Angularity by unit length (degrees/km)				
1 : 50 000 (crucial)[a]	81.282	71.774	154.625	171.175
1 : 200 000 (crucial)[a]	43.340	49.652	90.460	109.664
1 : 500 000 (crucial)[a]	6.856	41.820	91.112	57.145
1 : 200 000 automatically generalised[b]	68.077	59.642	125.473	148.031
1 : 500 000 automatically generalised[b]	51.565	56.292	125.074	131.930

[a] Features with only the crucial points, i.e. spurious points have been eliminated.
[b] The features at the scale of 1 : 50 000 have been generalised by the Douglas–Peucker algorithm until they have the same number of points as the same features at the scale of 1 : 200 000 and 1 : 500 000.

Table B.4 Comparing manual and automated generalisation for the Portuguese study area (displacement measures for the scales 1 : 50 000, 1 : 200 000 and 1 : 500 000).

	rd ac1 (245)	rail ac2	river ac32 (Rib. Grande)	bound ac7
Total areal displacement (m²)				
1 : 50 000 vs. 1 : 200 000	346 990.44	719 277.13	790 859.19	889 250.81
1 : 50 000 vs. 1 : 500 000	1 340 705.00	3 006 198.50	3 618 666.30	3 217 138.00
1 : 50 000 vs. 1 : 200 000 automatically generalised	10 600.26	32 652.04	30 644.17	25 788.53
1 : 50 000 vs. 1 : 500 000 automatically generalised	73 812.52	91 533.43	32 350.51	99 993.24
Total number of polygons				
1 : 50 000 vs. 1 : 200 000	12	11	37	22
1 : 50 000 vs. 1 : 500 000	7	4	11	12
1 : 50 000 vs. 1 : 200 000 automatically generalised	41	73	136	71
1 : 50 000 vs. 1 : 500 000 automatically generalised	28	48	136	64
Largest polygon (m²)				
1 : 50 000 vs. 1 : 200 000	122 164.48	413 110.62	123 221.96	196 489.16
1 : 50 000 vs. 1 : 500 000	449 788.50	2 956 923.70	879 054.50	1 259 666.80
1 : 50 000 vs. 1 : 200 000 automatically generalised	3250.37	2565.39	1094.33	2248.63
1 : 50 000 vs. 1 : 500 000 automatically generalised	10 752.79	8025.27	1094.33	10 095.69
Smallest polygon (m²)				
1 : 50 000 vs. 1 : 200 000	558.07	2809.49	0.05	62.11
1 : 50 000 vs. 1 : 500 000	1556.71	1185.25	91.45	13 718.85
1 : 50 000 vs. 1 : 200 000 automatically generalised	9.08	3.91	0.05	2.75
1 : 50 000 vs. 1 : 500 000 automatically generalised	7.46	31.29	0.05	3.97

Table B.4 *Continued*

	rd ac1 (245)	rail ac2	river ac32 (Rib. Grande)	bound ac7
Average polygon size (m^2)				
1 : 50 000 vs. 1 : 200 000	28 915.87	65 388.83	21 374.57	40 420.49
1 : 50 000 vs. 1 : 500 000	191 529.29	751 549.63	328 969.66	268 094.83
1 : 50 000 vs. 1 : 200 000 automatically generalised	258.54	447.29	225.32	363.22
1 : 50 000 vs. 1 : 500 000 automatically generalised	2636.16	1906.95	237.87	1562.39
Areal displacement per unit length (m^2/m)				
1 : 50 000 vs. 1 : 200 000	33.16	50.82	36.15	54.77
1 : 50 000 vs. 1 : 500 000	128.14	212.39	165.39	198.13
1 : 50 000 vs. 1 : 200 000 automatically generalised	1.01	2.31	1.40	1.59
1 : 50 000 vs. 1 : 500 000 automatically generalised	7.05	6.47	1.48	6.16
Number of polygons per unit length (polygons/km)				
1 : 50 000 vs. 1 : 200 000	1.1	0.8	1.7	1.4
1 : 50 000 vs. 1 : 500 000	0.7	0.3	0.5	0.7
1 : 50 000 vs. 1 : 200 000 automatically generalised	3.9	5.2	6.2	4.4
1 : 50 000 vs. 1 : 500 000 automatically generalised	2.7	3.4	6.2	3.9

Table B.5(a) Length and sinuosity for the British data (scales 1 : 50 000, 1 : 250 000 and 1 : 625 000).

	rd ac4 (B2068)	rd ac6 (B2077)	rd ac8 (A25)	rd ac9 (A2)	rd ac11 (M2)	rd ac 12 (A252)	rd ac13 (A28)	rd ac17 (A2050)	rd ac19 (A2050)	rd ac20 (A257)	rd ac21 (A290)	rd ac22 (A28)	rd ac23 (circ. rd)
Length (m)													
1 : 50 000	7625.8	984.3	10 717.8	19 116.6	4127.6	8927.1	12 780.0	3150.4	3581.1	4870.7	2771.7	3222.7	2784.3
1 : 250 000	7626.2	851.2	10 756.2	19 012.6	4163.1	8862.5	12 640.5	2994.3	3021.0	4962.3	2787.9	3403.0	2729.7
1 : 625 000	7091.2	734.2	10 756.9	19 122.0	4102.2	8228.0	11 733.2	1408.3	1966.5	5073.0	2580.9	3560.3	4641.6
Number of points (=n)													
1 : 50 000	65	40	282	270	56	234	232	64	35	52	56	30	77
1 : 250 000	34	5	63	87	19	53	68	18	12	22	19	16	20
1 : 625 000	11	3	24	29	9	11	21	3	5	9	4	7	11
Average segment length ($= length/(n - 1)$ (m))													
1 : 50 000	119.15	25.24	38.14	71.07	75.05	38.31	55.32	50.01	105.33	95.50	50.40	111.13	36.64
1 : 250 000	231.10	212.81	173.49	221.08	231.28	170.43	188.66	176.13	274.64	236.30	154.88	226.87	143.67
1 : 625 000	709.12	367.10	467.69	690.07	512.78	822.80	586.66	704.13	491.64	634.13	860.31	593.39	464.16
Cumulative angularity (degrees)													
1 : 50 000	374.65	151.20	1035.53	558.71	111.29	827.80	777.26	210.21	76.45	292.97	146.59	271.23	398.81
1 : 250 000	436.94	53.08	553.99	357.73	115.28	639.27	594.50	144.55	30.76	205.98	175.54	232.32	379.68
1 : 625 000	85.85	3.37	298.06	307.27	98.92	77.14	306.18	4.28	5.08	165.50	16.46	27.84	390.43
Angularity per unit length (degrees/km)													
1 : 50 000	49.13	153.61	96.62	28.18	26.96	92.73	60.82	66.72	21.35	60.15	52.89	84.16	143.24
1 : 250 000	57.29	62.36	51.50	23.92	27.69	72.13	47.03	48.28	10.18	41.51	62.97	68.27	139.09
1 : 625 000	12.11	4.59	27.71	15.90	24.11	9.38	26.10	3.04	2.58	32.62	6.38	7.82	84.11
Average angle ($= cumulative\ angularity/(n - 2)$ (degrees))													
1 : 50 000	5.95	3.98	3.70	2.01	2.06	3.57	3.38	3.39	2.32	5.86	2.71	9.69	5.32
1 : 250 000	13.65	17.69	9.08	4.68	6.78	12.53	9.01	9.03	3.08	10.30	10.33	16.59	21.09
1 : 625 000	9.54	3.37	13.55	11.38	14.13	8.57	16.11	4.28	1.69	23.64	8.23	5.57	43.38

Roads

Table B.5(b)   Length and sinuosity for the British data (continued).

	Railways		River	Boundaries		
	rail ac2 Ramsgate	rail ac3 Dover	riv ac24 G. Stour	bd ac26	bd ac27	bd ac28
Length (m)						
1 : 50 000	14 886.4	19 852.3	17 754.5	8409.5	5258.0	11 391.5
1 : 250 000	14 852.7	19 917.5	17 051.4			
1 : 625 000	15 004.3	20 204.9	16 179.8	7920.4	5802.1	10 719.1
Number of points (=n)						
1 : 50 000	437	413	444	197	64	170
1 : 250 000	69	81	123			
1 : 625 000	16	27	45	24	12	36
Average segment length (=length/(n − 1) (m))						
1 : 50 000	34.14	48.19	40.08	42.91	83.46	67.41
1 : 250 000	218.42	248.97	139.77			
1 : 625 000	1000.29	777.11	367.72	344.36	527.47	306.26
Cumulative angularity (degrees)						
1 : 50 000	370.68	595.24	6000.01	1588.61	934.14	1776.51
1 : 250 000	462.72	440.62	1939.49			
1 : 625 000	264.67	423.76	713.79	812.53	373.00	844.16
Angularity per unit length (degrees/km)						
1 : 50 000	24.90	29.98	337.94	188.91	177.66	155.95
1 : 250 000	31.15	22.12	113.74			
1 : 625 000	17.64	20.97	44.12	102.59	64.29	78.75
Average angle (=cumulative angularity/(n − 2) (degrees))						
1 : 50 000	0.85	1.45	13.57	8.15	15.07	10.57
1 : 250 000	6.91	5.58	16.03	0.00	0.00	0.00
1 : 625 000	18.90	16.95	16.60	36.93	37.30	24.83

Table B.6(a) Displacement measures for the British data (scales 1 : 50 000, 1 : 250 000 and 1 : 625 000).

						Roads							
	rd ac4 (B2068)	rd ac6 (B2077)	rd ac8 (A25)	rd ac9 (A2)	rd ac11 (M2)	rd ac12 (A252)	rd ac13 (A28)	rd ac17 (A2050)	rd ac19 (A2050)	rd ac20 (A257)	rd ac21 (A290)	rd ac22 (A28)	rd ac23 (circ. rd)
Total areal displacement (m²)													
50k–250k	501 360	46 980	577 750	2 877 652	194 646	320 537	565 742	242 927	272 326	351 010	390 768	70 127	162 483
50k–625k	1 313 510	75 485	1 655 256	6 042 448	420 308	1 325 435	1 014 993	1 594 555	715 412	835 459	604 572	922 827	1 486 845
Number of polygons													
50k–250k	4	2	5	5	4	14	7	3	1	2	1	3	4
50k–625k	4	2	4	5	3	2	13	1	1	3	1	1	2
Largest polygon (m²)													
50k–250k	269 649	46 705	362 796	2 260 628	163 940	92 114	185 680	127 978	272 326	340 633	390 768	42 942	95 566
50k–625k	1 120 096	41 246	660 365	5 063 520	256 709	1 303 509	231 208	1 594 555	715 412	646 097	604 572	922 827	1 068 076
Smallest polygon (m²)													
50k–250k	43	275	34	10	468	166	169	36 403	272 326	10 377	390 768	10 163	753
50k–625k	8 542	34 239	25 210	25 911	71 223	21 926	798	1 594 555	715 412	10 281	604 572	922 827	418 769
Average polygon size (= total areal displacement/no. polygons) (m²)													
50k–250k	125 340	23 490	115 550	575 530	48 661	22 895	80 820	80 976	272 326	175 505	390 768	23 376	40 621
50k–625k	328 377	37 743	413 814	1 208 490	140 103	662 717	78 076	1 594 555	715 412	278 486	604 572	922 827	743 423
Maximum vector displacement (m)													
50k–250k	122	102	110	323	92	110	97	191	144	143	170	59	140
50k–625k	426	142	328	864	191	383	200	927	290	399	285	297	593
Total areal displacement per unit length of the original line (m²/m)													
50k–250k	65.7	47.7	53.9	150.5	47.2	35.9	44.3	77.1	76.0	72.1	141.0	21.7	58.3
50k–625k	172.2	76.7	154.4	316.1	101.8	148.5	79.4	506.1	199.8	171.5	218.1	285.8	533.8
Number of polygons per unit length of the original line (polygons/km)													
50k–250k	0.5	2.0	0.5	0.3	1.0	1.6	0.5	1.0	0.3	0.4	0.4	0.9	1.4
50k–625k	0.5	2.0	0.4	0.3	0.7	0.2	1.0	0.3	0.3	0.6	0.4	0.3	0.7

Table B.6(b) Displacement measures for the British data (continued).

	Railways		River	Boundaries		
	rail ac2 Ramsgate	rail ac3 Dover	riv ac24 G. Stour	bd ac26	bd ac27	bd ac28
Total areal displacement (m^2)						
50k–250k	782 275	1 389 800	1 316 542			
50k–625k	1 567 143	4 302 451	3 195 526	1 610 288	1 637 948	3 951 156
Number of polygons						
50k–250k	7	9	12			
50k–625k	5	5	5	9	2	5
Largest polygon (m^2)						
50k–250k	431 424	575 068	548 778			
50k–625k	985 174	2 718 187	2 048 409	1 000 314	1 240 929	2 423 534
Smallest polygon (m^2)						
50k–250k	1989	103	32			
50k–625k	9709	1504	55 986	5257	397 019	483
Average polygon size (=*total areal displacement/no. polygons* (m^2))						
50k–250k	111 754	154 422	109 712			
50k–625k	313 429	860 490	639 105	178 921	818 974	790 231
Maximum vector displacement (m)						
50k–250k	126	144	256			
50k–625k	237	701	488	413	663	910
Total areal displacement per unit length of the original line (m^2/m)						
50k–250k	52.5	70.0	74.0			
50k–625k	105.3	216.7	179.7	191.5	311.5	346.9
Number of polygons per unit length of the original line (*polygons/km*)						
50k–250k	0.5	0.5	0.7			
50k–625k	0.3	0.3	0.3	1.1	0.4	0.4

Table B.7 Comparing manual and automated generalisation for the British study area (length and sinuosity) for the scales 1 : 50 000, 1 : 250 000 and 1 : 625 000.

	rd ac8 (A25)	rail ac2 (Ramsgate)	river ac24 (Great Stour)	bound ac26
Douglas–Peucker algorithm tolerance (DP) (m)				
1 : 50 000 (crucial)[a]	0.01	0.01	0.01	0.01
1 : 250 000 (crucial)[a]	0.1	0.1	0.1	
1 : 625 000 (crucial)[a]	1	1	1	1
1 : 250 000 automatically generalised[b]	6.3	2.85	9.55	
1 : 625 000 automatically generalised[b]	27.5	45.3	41	35
Number of points (= n)				
1 : 50 000 (crucial)[a]	254	405	441	194
1 : 250 000 (crucial)[a]	60	64	117	
1 : 625 000 (crucial)[a]	23	15	42	23
1 : 250 000 automatically generalised[b]	60	64	117	
1 : 625 000 automatically generalised[b]	23	15	42	23
Length (m)				
1 : 50 000 (crucial)[a]	10 717.770	14 886.378	17 754.529	8409.542
1 : 250 000 (crucial)[a]	10 756.201	14 852.739	17 051.387	
1 : 625 000 (crucial)[a]	10 756.865	15 004.319	16 179.765	7920.38
1 : 250 000 automatically generalised[b]	10 704.782	14 884.357	17 651.861	
1 : 625 000 automatically generalised[b]	10 660.355	14 851.264	17 342.082	8298.607

Table B.7 *Continued*

	rd ac8 (A25)	rail ac2 (Ramsgate)	river ac24 (Great Stour)	bound ac26
Average segment length (m)				
1 : 50 000 (crucial)[a]	42.363	36.847	40.351	43.573
1 : 250 000 (crucial)[a]	182.308	235.758	146.995	
1 : 625 000 (crucial)[a]	488.948	1071.737	394.628	360.018
1 : 250 000 automatically generalised[b]	181.437	236.260	152.171	
1 : 625 000 automatically generalised[b]	484.562	1060.805	422.978	377.209
Cumulative angularity (degrees)				
1 : 50 000 (crucial)[a]	1035.443	370.487	5999.892	1588.420
1 : 250 000 (crucial)[a]	553.021	462.566	1938.789	
1 : 625 000 (crucial)[a]	293.916	264.542	712.740	812.528
1 : 250 000 automatically generalised[b]	697.659	326.409	3088.980	
1 : 625 000 automatically generalised[b]	416.233	218.438	1476.612	788.865
Angularity by unit length (degrees/km)				
1 : 50 000 (crucial)[a]	96.610	24.888	337.936	188.883
1 : 250 000 (crucial)[a]	51.414	31.143	113.703	
1 : 625 000 (crucial)[a]	27.324	17.631	44.051	102.587
1 : 250 000 automatically generalised[b]	65.173	21.930	174.995	
1 : 625 000 automatically generalised[b]	39.045	14.708	85.146	95.060

[a] Features with only the crucial points, i.e. spurious points have been eliminated.
[b] The features at the scale of 1 : 50 000 have been generalised by the Douglas–Peucker algorithm until they have the same number of points as the same features at the scales of 1 : 250 000 and 1 : 625 000.

Table B.8 Comparing manual and automated generalisation for the British study area (displacement measures for the scales 1:50000, 1:250000 and 1:625000).

	rd ac8 (A25)	rail ac2 (Ramsgate)	river ac24 (Great Stour)	bound ac26
Total areal displacement (m²)				
1:50000 vs. 1:250000	577 749.56	782 275.13	1 316 541.90	
1:50000 vs. 1:625000	1 655 255.90	1 567 142.90	3 195 526.00	1 610 288.00
1:50000 vs. 1:250000 automatically generalised	14 819.79	8965.24	41 069.44	
1:50000 vs. 1:625000 automatically generalised	97 330.06	189 022.61	177 371.42	70 046.04
Total number of polygons				
1:50000 vs. 1:250000	5	7	12	
1:50000 vs. 1:625000	4	5	5	9
1:50000 vs. 1:250000 automatically generalised	85	59	188	
1:50000 vs. 1:625000 automatically generalised	41	22	102	36
Largest polygon (m²)				
1:50000 vs. 1:250000	362 796.47	431 424.31	548 778.37	
1:50000 vs. 1:625000	660 364.94	985 173.75	2 048 408.90	1 000 314.30
1:50000 vs. 1:250000 automatically generalised	865.92	982.25	1582.01	
1:50000 vs. 1:625000 automatically generalised	15 353.75	36 616.54	13 877.98	9368.17
Smallest polygon (m²)				
1:50000 vs. 1:250000	33.78	1989.36	32.23	
1:50000 vs. 1:625000	25 210.48	9708.73	55 986.22	5256.82
1:50000 vs. 1:250000 automatically generalised	0.63	1.04	0.98	
1:50000 vs. 1:625000 automatically generalised	1.13	16.10	0.59	2.19

Table B.8 *Continued*

	rd ac8 (A25)	rail ac2 (Ramsgate)	river ac24 (Great Stour)	bound ac26
Average polygon size (m²)				
1 : 50 000 vs. 1 : 250 000	115 549.91	111 753.59	109 711.83	
1 : 50 000 vs. 1 : 625 000	413 813.98	313 428.58	639 105.20	178 920.89
1 : 50 000 vs. 1 : 250 000 automatically generalised	174.35	151.95	218.45	
1 : 50 000 vs. 1 : 625 000 automatically generalised	2373.90	8591.94	1738.94	1945.72
Areal displacement per unit length (m²/m)				
1 : 50 000 vs. 1 : 250 000	53.91	52.55	74.15	
1 : 50 000 vs. 1 : 625 000	154.44	105.27	179.98	191.48
1 : 50 000 vs. 1 : 250 000 automatically generalised	1.38	0.60	2.31	
1 : 50 000 vs. 1 : 625 000 automatically generalised	9.08	12.70	9.99	8.33
Number of polygons per unit length (polygons/km)				
1 : 50 000 vs. 1 : 250 000	0.5	0.5	0.7	
1 : 50 000 vs. 1 : 625 000	0.4	0.3	0.3	1.1
1 : 50 000 vs. 1 : 250 000 automatically generalised	7.9	4.0	10.6	
1 : 50 000 vs. 1 : 625 000 automatically generalised	3.8	1.5	5.7	4.3

Table B.9 British data – 1 : 10 000 scale versus 1 : 50 000 scale.

	Roads					Railways		Rivers	
Scales	rd ac9 (A2)	rd ac13 (A28)	rd ac17 (A2050)	rd ac21 (A290)	rd ac23 (circular rd)	rail ac2 (to Ramsgate)	rail ac3 (to Dover)	rv ac24 (G. Stour)	rv ac25 (G. Stour)
Length (m)									
1:10000	6097.24	5283.05	3013.09	2774.33	1794.25	6425.84	5691.94	5782.79	668.49
1:50000	6098.90	5307.36	3150.40	2770.42	1736.75	6397.46	5701.18	5745.34	611.44
Number of points (=n)									
1:10000	102	148	85	91	60	174	161	255	29
1:50000	52	107	64	56	47	304	155	124	12
Average segment length (=length/(n − 1)) (m)									
1:10000	60.37	35.94	35.87	30.83	30.41	37.14	35.57	22.77	23.87
1:50000	119.59	50.07	50.01	50.37	37.76	21.11	37.02	46.71	55.59
Cumulative angularity (degrees)									
1:10000	224.42	711.44	249.54	417.81	482.97	530.45	541.60	1800.05	148.78
1:50000	63.63	449.71	210.21	146.59	239.65	164.24	254.72	1378.75	67.26
Angularity per unit length (degrees/km)									
1:10000	36.81	134.66	82.82	150.60	269.18	82.55	95.15	311.28	222.56
1:50000	10.43	84.73	66.72	52.91	137.99	25.67	44.68	239.98	110.00
Average angle (=cumulative angularity/(n − 2)) (degrees)									
1:10000	2.24	4.87	3.01	4.69	8.33	3.08	3.41	7.11	5.51
1:50000	1.27	4.28	3.39	2.71	5.33	0.54	1.66	11.30	6.73

Table B.9 *Continued*

Scales	Roads					Railways		Rivers	
	rd ac9 (A2)	rd ac13 (A28)	rd ac17 (A2050)	rd ac21 (A290)	rd ac23 (circular rd)	rail ac2 (to Ramsgate)	rail ac3 (to Dover)	rv ac24 (G. Stour)	rv ac25 (G. Stour)
Müller (1987) angularity measure									
1:10 000	1.0105	1.0097	1.0130	1.0150	1.0278	1.0069	1.0077	1.0098	1.0402
1:50 000	1.0202	1.0124	1.0179	1.0195	1.0258	1.0034	1.0071	1.0245	1.1039
Total areal displacement (m²)									
10k–50k	9136.0	37 412.6	40 756.9	19 258.3	20 030.1	38 177.5	26 040.2	46 679.6	18 894.9
Number of polygons									
10k–50k	27	18	3	11	7	13	16	35	1
Largest polygon (m²)									
10k–50k	1357.7	15 677.9	23 029.1	6262.1	9328.8	12 150.5	8986.5	10 640.0	18 894.9
Smallest polygon (m²)									
10k–50k	0.0	0.0	3586.2	4.3	78.3	16.4	0.0	2.2	18 894.9
Average size of polygon (=total areal displacement/no. of polygons (m²))									
10k–50k	338.4	2078.5	13 585.6	1750.8	2861.4	2936.7	1627.5	1333.7	18 894.9
Maximum vector displacement (m)									
10k–50k	4.2	27.6	25.2	18.2	29.8	17.0	14.1	27.5	34.6
Total areal displacement per unit length (m²/m)									
10k–50k	1.5	7.1	13.5	6.9	11.2	5.9	4.6	8.1	28.3
Number of polygons per unit length (polygons/km)									
10k–50k	4.4	3.4	1.0	4.0	3.9	2.0	2.8	6.1	1.5

Sample monochrome copies of the maps used

In this appendix, sample copies of the six maps used are presented in the following order: for the Portuguese study area, Figures C.1–C.3, and for the British study area, Figures C.4–C.6.

Figures C.4–C.6 are reproduced from Ordnance Survey maps with the permission of The Controller of Her Majesty's Stationery Office, © Crown copyright. ED 275484.

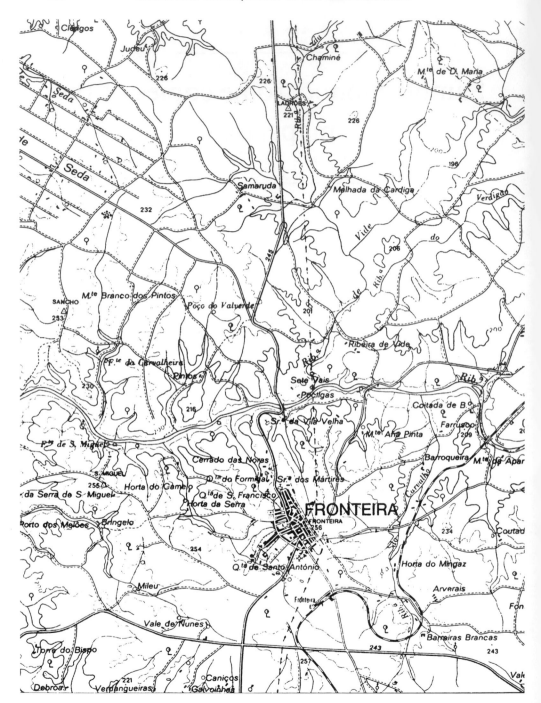

Figure C.1   Part of sheet 32-D, Sousel (1975), of the Instituto Português de Cartografia e Cadastro 1 : 50 000 scale map. © Instituto Português de Cartografia e Cadastro.

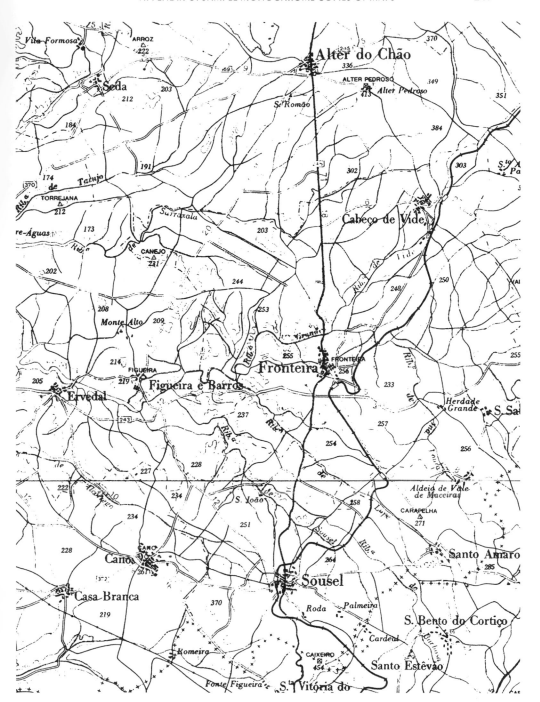

Figure C.2 Part of sheet 6 (1974) of the Instituto Português de Cartografia e Cadastro 1 : 200 000 scale map. © Instituto Português de Cartografia e Cadastro.

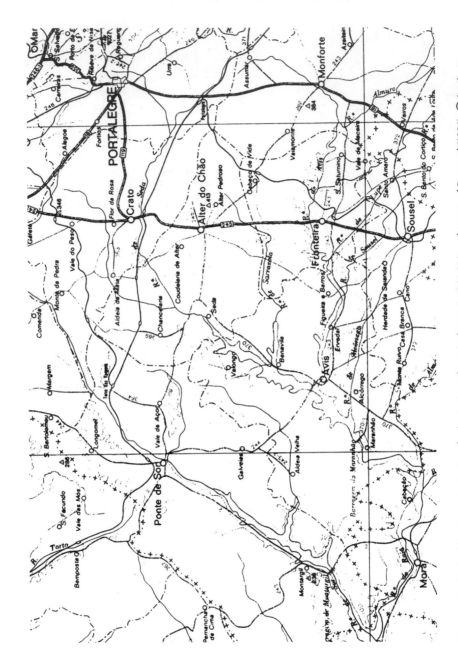

Figure C.3 Part of the Instituto Português de Cartografia e Cadastro 1 : 500 000 scale map of Portugal (1981). © Instituto Português de Cartografia e Cadastro.

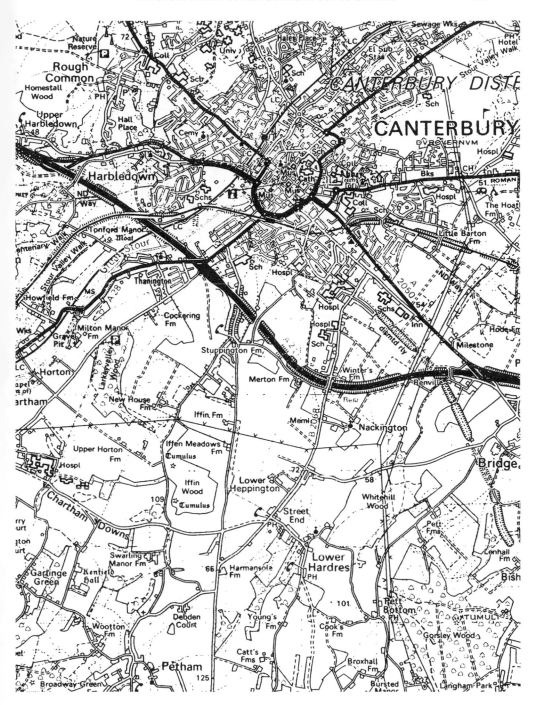

Figure C.4 Part of sheet 179, Canterbury & East Kent (1993), of the Ordnance Survey 1 : 50 000 scale map (Landranger series). © Crown copyright.

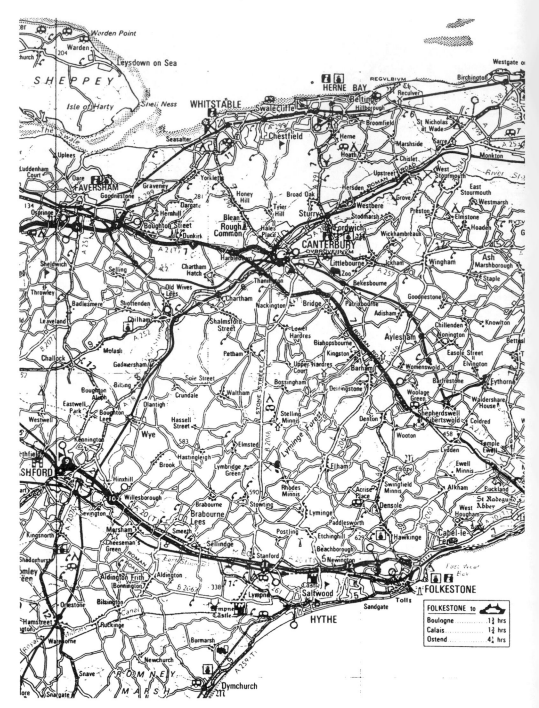

Figure C.5   Part of sheet 9, South-East England (1984), of the Ordnance Survey 1 : 250 000 scale map (Routemaster series). © Crown copyright.

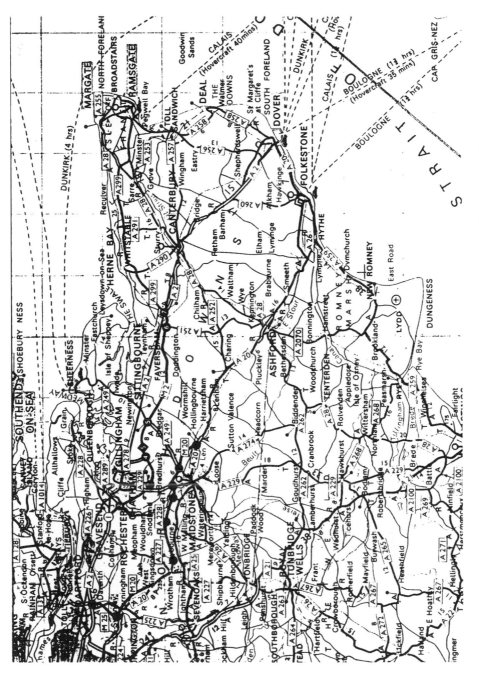

Figure C.6 Part of the South sheet (1977) of the Ordnance Survey 1 : 625 000 scale map. © Crown copyright.

References

ANSELIN, L., 1989, What is special about spatial data? Alternative perspectives on spatial data analysis, NCGIA Technical Paper 89-4, June.

ARONOFF, S., 1989, *Geographic Information Systems: A Management Perspective*, Ottawa: WDL Publications.

AUSTRALIAN SURVEY, 1992, *TOPO-250k Data User Guide*, Belconnen: Australian Surveying and Land Information Group.

AVILES, C., SCHOLZ, C. and BOATWRIGHT, J., 1987, Fractal analysis applied to characteristic segments of the San Andreas fault, *Journal of Geophysical Research*, **92**(B1), 331–44.

BAEIJS, C., DEMAZEAU, Y. and ALVARES, L., 1996, SIGMA: application of multi-agent systems to cartographic generalization, *7th European Workshop on Modelling Autonomous Agents in a Multi-agent World (MAAMAW '96)*, Eindhoven, The Netherlands, Lecture Notes in Artificial Intelligence 1038, Berlin: Springer-Verlag.

BAELLA, B., COLOMER, J. and PLA, M., 1994, CHANGE: Technical report, Barcelona: Institut Cartogràfic de Catalunya.

BALODIS, M., 1988, Generalisation, in ANSON, R. (Ed.), *Basic Cartography for Students and Technicians*, Vol. 2, International Cartographic Association, London: Elsevier Applied Science, pp. 71–84.

BATTY, M., FOTHERINGAM, S. and LONGLEY, P., 1989, Urban growth and form: scaling, fractal geometry and diffusion limited aggregation, *Environment and Planning A*, **21**(11), 1447–72.

BEARD, K., 1987, How to survive on a single detailed database, *Proceedings of AUTO-CARTO 8*, Baltimore, Maryland, 30 March – 2 April, pp. 211–20.

BEARD, K., 1988, Multiple representations from a detailed database: a scheme for automated generalization, Unpublished PhD thesis, The University of Wisconsin–Madison.

BEARD, K., 1989, User error: the neglected error component, *Proceedings of AUTO-*

CARTO 9, Baltimore, Maryland, 2–7 April, pp. 808–17.

BEARD, K., 1991a, Constraints on rule formation, in BUTTENFIELD, B. and MCMASTER, R. (Eds.), *Map Generalization: Making Rules for Knowledge Representation*, Harlow: Longman, pp. 121–35.

BEARD, K., 1991b, User interaction in map generalisation, *15th Conference of the International Cartographic Association, Special Session on Generalisation*, 28 September, Bournemouth, England.

BLAKEMORE, M., 1983, Generalization and error in spatial data bases, *Proceedings of AUTO-CARTO 6*, Falls Church, Virginia, pp. 313–22.

BLAKEMORE, M., 1985, High or low resolution? Conflicts of accuracy, cost, quality, and application in computer mapping, *Computers & Geosciences*, **11**(3), 345–8.

BOLSTAD, P., GESSLER, P. and LILLESAND, T., 1990, Positional uncertainty in manually digitised map data, *International Journal of Geographical Information Systems*, **4**(4) 399–412.

BRASSEL, K. and WEIBEL, R., 1988, A review and framework of automated map generalisation, *International Journal of Geographical Information Systems*, **2**(3) 229–44.

BRUNSDON, C. and OPENSHAW, S., 1993, Simulating the effects of error in GIS, in MATHER, P. (Ed.), *Geographical Information Handling – Research and Applications*, Chichester: John Wiley, pp. 47–61.

BUNDY, G., JONES, C. and FURSE, E., 1995, Holistic generalization of large-scale cartographic data, in MÜLLER, J.-C., LAGRANGE, J.-P. and WEIBEL, R. (Eds.), *GIS and Generalisation: Methodology and Practice*, London: Taylor & Francis, pp. 106–19.

BURROUGH, P., 1983, Multi-scale sources of spatial variation in soil. I. The application of fractal concepts to nested levels of soil variation, *Journal of Soil Science*, **34**, 577–97.

BUTTENFIELD, B., 1986, Digital definitions of scale-dependent structure, *Proceedings of AUTO-CARTO London*, Vol. 1, pp. 497–506.

BUTTENFIELD, B., 1989, Scale-dependence and self-similarity in cartographic lines, *Cartographica*, **26**(1) 79–100.

BUTTENFIELD, B. and MCMASTER, R. (Eds.), 1991, *Map Generalization: Making Rules for Knowledge Representation*, Harlow: Longman.

BUTTENFIELD, B., WEBER, C., LEITNER, M., PHELAN, J., RASMUSSEN, D. and WRIGHT, G., 1991, How does a cartographic object behave? Computer inventory of topographic maps, *Proceedings of GIS/LIS '91*, Atlanta, Georgia, 28–31 October, Vol. 2, pp. 891–900.

CARSTENSEN, L., 1990, Angularity and capture of the cartographic line during digital data entry, *Cartography and Geographic Information Systems*, **17**(3), 209–24.

CATLOW, D. and DAOSHENG, D., 1984, The structuring and generalisation of digital river data, *ACSM 44th Annual Meeting, Technical Papers*, Washington, DC, pp. 511–20.

CHANG, K. and TSAI, B., 1989, The effect of DEM resolution on spatial interpolation, *Proceedings of GIS/LIS '89*, Orlando, Florida, 26–30 November, pp. 40–9.

CHRISMAN, N., 1987, Obtaining information on quality of digital data, *Proceedings of AUTO-CARTO 8*, Baltimore, Maryland, 30 March – 2 April, pp. 350–8.

CHRISMAN, N., 1989, Error in categorical maps: testing versus simulation, *Pro-

ceedings of AUTO-CARTO 9, Baltimore, Maryland, 2–7 April, pp. 521–9.

CHRISMAN, N., 1991, The error component in spatial data, in MAGUIRE, D., GOODCHILD, M. and RHIND, D. (Eds.), *Geographical Information Systems: Principles and Applications*, Vol. 1, Harlow: Longman, pp. 165–74.

CROMLEY, R., 1992a, *Digital Cartography*, Englewood Cliffs, New Jersey: Prentice Hall.

CROMLEY, R., 1992b, Principal axis line simplification, *Computers & Geosciences*, **18**(8) 1003–11.

CROSS, A., 1989, Childhood leukaemia in northern England, Unpublished PhD thesis, Department of Geography, University of Newcastle upon Tyne.

DEPARTMENT OF THE ENVIRONMENT, 1987, *Handling Geographic Information*, Report of the Committee of Enquiry chaired by Lord Chorley, London: HMSO.

DEVEREUX, B. and MAYO, T., 1990, Intelligent techniques for cartographic data capture, *Proceedings of AGI '90*, Birmingham, pp. 6.2.1–9.

DEVOGELE, T., TREVISA, J. and RAYNAL, L., 1997, Building a multi-scale database with scale-transition relationships, in KRAAK, M.-J., MOLENAAR, M. and FENDEL, E. (Eds.), *Advances in GIS Research II* (Proceedings of the 7th International Symposium on Spatial Data Handling), London: Taylor & Francis, pp. 337–51.

DOUGLAS, D. and PEUCKER, T., 1973, Algorithms for the reduction of the number of points required to represent a digitised line or its caricature, *Canadian Cartographer*, **10**, 112–22.

DUTTON, G., 1981, Fractal enhancement of cartographic line detail, *The American Cartographer*, **8**(1), 23–40.

ECKERT, M., 1908, On the nature of maps and map logic (translated by W. Joerg), *Bulletin of the American Geographical Society*, **40**(6), 344–51.

EGENHOFER, M. and FRANK, A., 1989, Object-oriented modelling in GIS: inheritance and propagation, *Proceedings of AUTO-CARTO 9*, Baltimore, Maryland, 2–7 April, pp. 588–98.

EHRLIHOLZER, R., 1995, Quality assessment in generalization: integrating quantitative and qualitative methods, *Proceedings of the 17th International Cartographic Conference: 'Cartography Crossing Borders'*, Barcelona: Institut Cartogràfic de Catalunya, pp. 2241–50.

FELLOWS, J. and RAGAN, R., 1986, The role of cell size in hydrology oriented GIS, *Hydrologic Applications of Space Technology* (Proceedings of the Cocoa Beach Workshop, Florida, August 1985). IAHS Publ. no. 160.

FISHER, P., 1989, Knowledge-based approaches to determining and correcting areas of unreliability in geographic databases, in GOODCHILD, M. and GOPAL, S. (Eds.), *Accuracy of Spatial Databases*, London: Taylor & Francis, pp. 45–54.

FISHER, P., 1991, Spatial data sources and data problems, in MAGUIRE, D., GOODCHILD, M. and RHIND, D. (Eds.), *Geographical Information Systems: Principles and Applications*, Vol. 1, Harlow: Longman, pp. 175–89.

FISHER, P. and MACKANESS, W., 1987, Are cartographic expert systems possible? *Proceedings of AUTO-CARTO 8*, Baltimore, Maryland, 30 March – 2 April, pp. 530–4.

FRANK, A., 1991, Design of cartographic databases, in MÜLLER, J.-C. (Ed.), *Advances in Cartography*, International Cartographic Association, London: Elsevier Applied Science, pp. 15–44.

FRITSCH, E. and LAGRANGE, J.-P., 1995, Spectral representations of linear features for generalisation, in FRANK, A. and KUHN, W. (Eds.), *Proceedings of COSIT '95: 'Spatial Information Theory – a Theoretical Basis for GIS'*, Lecture Notes in Computer Science 988. Berlin: Springer-Verlag, pp. 157–71.

GALLOWAY, R. and BAHR, M., 1979, What is the length of the Australian coast? *Australian Geographer*, **14**, 244–7.

GARDINER, V., 1982, Stream networks and digital cartography, in BICKMORE, D. (Ed.), Monograph 28, 'Perspectives in the alternative cartography – cartographic computing technology and applications', *Cartographica*, **19**(2), 38–44.

GEHLKE, C. and BIEHL, K., 1934, Certain effects of grouping upon the size of the correlation coefficient in census tract material, *Journal of the American Statistical Association Supplement*, **29**, 169–70.

GOODCHILD, M., 1980a, Fractals and the accuracy of geographic measures, *International Association for Mathematical Geology Journal*, **12**(2), 85–98.

GOODCHILD, M., 1980b, The effects of generalisation in geographical data encoding, in FREEMAN, H. and PIERONI, G. (Eds.), *Map Data Processing*, New York: Academic Press, pp. 191–205.

GOODCHILD, M., 1991, Issues of quality and uncertainty, in MÜLLER, J.-C. (Ed.), *Advances in Cartography*, International Cartographic Association, London: Taylor & Francis, pp. 113–39.

GOODCHILD, M. and GOPAL, S. (Eds.), 1989, *Accuracy of Spatial Databases*, London: Taylor & Francis.

GOODCHILD, M. and MARK, D., 1987, The fractal nature of geographic phenomena, *Annals of the Association of American Geographers*, **72**(2), 265–78.

GRÜNREICH, D., 1993, Generalisation in GIS environment, *Proceedings of the 16th International Cartographic Conference*, Cologne, pp. 203–10.

GRÜNREICH, D., POWITZ, B. and SCHMIDT, C., 1992, Research and development in computer-assisted generalisation of topographic information at the Institute of Cartography, Hannover University, *Proceedings of the Third European Conference on Geographical Information Systems*. Münich, Germany, 23–26 March, pp. 532–41.

GUEDES, J., 1988, *Automatização na Cartografia Básica e Derivada*, Lisboa: Instituto Português de Cartografia e Cadastro.

GUPTILL, S., 1989, Speculations on seamless, scaleless cartographic data bases, *Proceedings of AUTO-CARTO 9*, Baltimore, Maryland, 2–7 April, pp. 436–43.

HÅKONSON, L., 1978, The length of closed geomorphic lines, *Mathematical Geology*, **10**(2), 141–67.

HARLEY, J., 1975, *Ordnance Survey Maps: a Descriptive Manual*, Southampton: Ordnance Survey.

HERBERT, G. and JOÃO, E., 1991, Automating map design and generalisation: a review of systems and prospects for future progress in the 1990s, South East Regional Research Laboratory Working Report No. 27, Birkbeck College, London.

HERBERT, G., JOÃO, E. and RHIND, D., 1992, Use of an artificial intelligence approach to increase user control of automatic line generalisation, *Proceedings of the Third European Conference on Geographical Information Systems*. Münich, Germany, 23–26 March, pp. 554–63.

HERSHBERGER, J. and SNOEYINK, J., 1992, Speeding up the Douglas–Peucker line-simplification algorithm, *Proceedings of the 5th International Symposium on Spatial Data Handling*, Charleston, South Carolina, 3–7 August, Vol. 1, pp. 134–43.

IMHOF, E., 1982, *Cartographic Relief Presentation*, Berlin: Walter de Gruyter.

INSTITUTO PORTUGUÊS DE CARTOGRAFIA E CADASTRO, 1988, *Informação sobre Documentação e Elementos de Estudo Disponiveis*, Lisboa: Instituto Português de Cartografia e Cadastro.

JASINSKI, M., 1990, The comparison of complexity measures for cartographic lines, NCGIA Report no. 90-1. National Center for Geographical Information and Analysis, University of California, Santa Barbara.

JENKS, J., 1979, Thoughts on line generalisation, *Proceedings of AUTO-CARTO 4*, pp. 205–20.

JOÃO, E., 1991, The role of the user in generalisation within GIS, in MARK, D. and FRANK, A. (Eds.), *Cognitive and Linguistic Aspects of Geographic Space*, Dordrecht: Kluwer Academic, pp. 493–506.

JOÃO, E., RHIND, D., OPENSHAW, S. and KELK, B., 1990, Generalisation and GIS databases, *Proceedings of the First European Conference on Geographical Information Systems*, Amsterdam, The Netherlands, 10–13 April, pp. 504–16.

JOÃO, E., HERBERT, G., RHIND, D., OPENSHAW, S. and RAPER, J., 1993, Towards a generalisation machine to minimise generalisation effects within a GIS, in MATHER, P. (Ed.), *Geographical Information Handling – Research and Applications*, Chichester: John Wiley, pp. 63–78.

JONES, C. and ABRAHAM, I., 1986, Design considerations for a scale-independent cartographic data base, *Proceedings of the 2nd International Symposium on Spatial Data Handling*, Seattle, Washington, August, pp. 384–98.

KEATES, J., 1989, *Cartographic Design and Production*, Harlow: Longman.

KELLER, S., 1995, 'Generalization by example': interactive parameter control in line generalization using genetic algorithms, *Proceedings of the 17th International Cartographic Conference: 'Cartography Crossing Borders'*, Barcelona: Institut Cartogràfic de Catalunya, pp. 1974–83.

KLINKENBERG, B. and GOODCHILD, M., 1992, The fractal properties of topography: a comparison of methods, *Earth Surface Processes and Landforms*, **17**, 217–34.

KRANTZ, S., 1989, Fractal geometry, *The Mathematical Intelligencer*, **11**(4), 12–16.

LANGRAN, G., 1992, *Time in Geographic Information Systems*, London: Taylor & Francis.

LASER-SCAN, 1992, *Laser-Scan User's Guide*, Cambridge: Laser-Scan, Ltd.

LASER-SCAN, 1997, *LAMPS2 Object Processing Guide 1.0*, February 1997, Cambridge: Laser-Scan, Ltd.

LEE, D., 1992, Cartographic generalization – from large scale to smaller scale, Unpublished technical report describing the design aspects incorporated into the Intergraph's 'MGE Map Generalizer', Alabama: Intergraph Corporation.

LEE, D., 1995, Experiment on formalizing the generalization process, in MÜLLER, J.-C., LAGRANGE, J.-P. and WEIBEL, R. (Eds.), *GIS and Generalisation: Methodology and Practice*, London: Taylor & Francis, pp. 219–34.

LEITNER, M. and BUTTENFIELD, B., 1995, Acquisition of procedural cartographic knowledge by reverse engineering, *Cartography and Geographic Information Systems*, **22**(3), 232–41.

LI, Z. and OPENSHAW, S., 1992, Algorithms for automated line generalisation based on a natural principle of objective generalisation, *International Journal of Geographical Information Systems*, **6**(5), 373–89.

LI, Z. and OPENSHAW, S., 1993, A natural principle for the objective generalisation of digital maps, *Cartography and Geographic Information Systems*, **20**(1), 19–29.

LIEBSCHER, H., 1982, Introduction to river networks and computer mapping, in BICKMORE, D. (Ed.), Monograph 28, 'Perspectives in the alternative cartography – cartographic computing technology and applications', *Cartographica*, **19**(2), 36–7.

MACKANESS, W., 1994, An algorithm for conflict identification and feature displacement in automated map generalization, *Cartography and Geographic Information Systems*, **21**(4), 219–32.

MACKANESS, W. and BEARD, K., 1990, Development of an interface for user interaction in rule based map generalisation, *Proceedings of GIS/LIS '90*, Anaheim, California, pp. 107–16.

MACKANESS, W. and FISHER, P., 1987, Automatic recognition and resolution of spatial conflicts in cartographic symbolisation, *Proceedings of AUTO-CARTO 8*, Baltimore, Maryland, 30 March–2 April, pp. 709–18.

MACKANESS, W. and SCOTT, D., 1988, The problems of operationally defining the map design process for cartographic expert systems, *Proceedings of Austra Carto III, 7th Australian Cartographic Conference*, Sydney, Australia, pp. 715–23.

MALING, D., 1989, *Measurements from Maps – Principles and Methods of Cartometry*, Oxford: Pergamon Press.

MANDELBROT, B., 1967, How long is the coast of Britain? *Science*, **156**, 636–8.

MANDELBROT, B., 1982, *The Fractal Geometry of Nature*, San Francisco: W. H. Freeman.

MARINO, J., 1979, Identification of characteristic points along naturally occurring lines: an empirical study, *The Canadian Cartographer*, **16**(1), 70–80.

MARK, D., 1990, Competition for map space as a paradigm for automated map design, *Proceedings of GIS/LIS '90*, Anaheim, California, pp. 97–106.

MARTIN, C., 1988, *User-centered Requirements Analysis*, Englewood Cliffs, New Jersey: Prentice Hall.

MASSER, I., 1990, GIS in Britain: the Regional Research Laboratory Initiative, *Proceedings of the First European Conference on Geographical Information Systems*, Amsterdam, The Netherlands, 10–13 April, Vol. 2, pp. 721–8.

MCHARG, I., 1969, *Design with Nature*, New York: Natural History Press.

MCMASTER, R., 1986, A statistical analysis of mathematical measures for linear simplification, *The American Cartographer*, **13**(2), 103–16.

MCMASTER, R., 1987, The geometric properties of numerical generalisation, *Geographical Analysis*, **19**(4), 330–46.

MCMASTER, R., 1989, Introduction to 'Numerical generalization in cartography', in MCMASTER, R. (Ed.), Monograph 40, 'Numerical generalization in cartography', *Cartographica*, **26**(1), 1–6.

MCMASTER, R., 1991, Conceptual frameworks for geographical knowledge, in BUTTENFIELD, B. and MCMASTER, R. (Eds.), *Map Generalization: Making Rules for Knowledge Representation*, Harlow: Longman, pp. 21–39.

MCMASTER, R. and MONMONIER, M., 1989, A conceptual framework for quantitative and qualitative raster-mode generalisation, *Proceedings of GIS/LIS '89*, Orlando, Florida, 26–30 November, Vol. 2, pp. 390–403.

MOELLERING, H. and TOBLER, W., 1972, Geographical variances, *Geographical Analysis*, **4**, 34–50.

MONMONIER, M., 1982, *Computer-Assisted Cartography – Principles and Prospects*, Englewood Cliffs, New Jersey: Prentice Hall.

MONMONIER, M., 1983, Raster-mode area generalisation for land use and land cover maps, *Cartographica*, **20**(4), 65–91.

MONMONIER, M., 1986, Towards a practicable model of cartographic generalisation, *Proceedings of AUTO-CARTO London*, London, Vol. 2, pp. 257–66.

MONMONIER, M., 1987, Displacement in vector- and raster-mode graphics. *Cartographica*, **24**(4), 25–36.

MONMONIER, M., 1989, Regionalizing and matching features for interpolated displacement in the automated generalisation of digital cartographic databases, *Cartographica*, **26**(2), 21–39.

MONMONIER, M., 1991, *How to Lie with Maps*, Chicago: The University of Chicago Press.

MÜLLER, J.-C., 1986, Fractal dimension and inconsistencies in cartographic line representations, *The Cartographic Journal*, **23**, 123–30.

MÜLLER, J.-C., 1987, Fractal and automatic line generalisation, *The Cartographic Journal*, **24**, 27–34.

MÜLLER, J.-C., 1989, Theoretical considerations for automated map generalisation, *ITC Journal*, **3/4**, 200–4.

MÜLLER, J.-C., 1990a, Rule based generalization: potentials and impediments, *Proceedings of the International Symposium on Spatial Data Handling*, Zürich, Vol. 1, pp. 317–34.

MÜLLER, J.-C., 1990b, The removal of spatial conflicts in line generalisation, *Cartography and Geographic Information Systems*, **17**(2), 141–9.

MÜLLER, J.-C., 1991a, Building knowledge tanks for rule based generalisation, *Proceedings of the ICA Conference*, Bournemouth, pp. 257–66.

MÜLLER, J.-C., 1991b, Generalization of spatial databases, in MAGUIRE, D., GOODCHILD, M. and RHIND, D. (Eds.), *Geographical Information Systems: Principles and Applications*, Vol. 1, Harlow: Longman, pp. 457–75.

MÜLLER, J.-C. and MOUWES, P., 1990, Knowledge acquisition and representation for rule-based map generalisation: an example from The Netherlands, *Proceedings of GIS/LIS '90*, Anaheim, California, pp. 58–67.

MÜLLER, J.-C. and WANG, Z., 1992, Area-patch generalisation: a competitive approach, *The Cartographic Journal*, **29**, 137–44.

MÜLLER, J.-C., WEIBEL, R., LAGRANGE, J.-P. and SALGÉ, F., 1995, Generalisation: state of the art and issues, in MÜLLER, J.-C., LAGRANGE, J.-P. and WEIBEL, R. (Eds.), *GIS and Generalisation: Methodology and Practice*, London: Taylor & Francis, pp. 3–17.

NCDCDS, 1988, The proposed standard for digital cartographic data – part III: digital cartographic data quality (NCDCDS – National Committee for Digital Cartographic Data Standards), *The American Cartographer*, **15**(1), 129–35.

NCGIA, 1989, *Technical Issues in GIS – NCGIA Core Curriculum*, National Center for Geographic Information and Analysis, University of California, Santa Barbara.

NEPRASH, J., 1934, Some problems in the correlation of spatially distributed variables, *Journal of the American Statistical Association Supplement*, **29**, 167–8.

OPENSHAW, S., 1984, *The Modifiable Areal Unit Problem*, CATMOG, Concepts and

Techniques in Modern Geography, no. 38, Norwich: Geo Abstracts.

OPENSHAW, S., 1989, Learning to live with errors in spatial databases, in GOOD-CHILD, M. and GOPAL, S. (Eds.), *Accuracy of Spatial Databases*, London: Taylor & Francis, pp. 263–76.

OPENSHAW, S., 1996, Developing GIS-relevant zone-based spatial analysis methods, in LONGLEY, P. and BATTY, M. (Eds.), *Spatial Analysis: Modelling in a GIS Environment*, Cambridge: GeoInformation International, pp. 55–73.

OPENSHAW, S., CHARLTON, M. and CARVER, S., 1991, Error propagation: a Monte Carlo simulation, in MASSER, I. and BLAKEMORE, M. (Eds.), *Handling Geographical Information: Methodology and Potential Applications*, Harlow: Longman, pp. 78–101.

ORDNANCE SURVEY, 1987, *Ordnance Survey 1 : 625 000 Digital Data User Manual*, Southampton: Ordnance Survey.

ORDNANCE SURVEY, 1989, Ordnance Survey Specification Module 7E for the 1 : 10 000 Scale Map, Internal Guidelines, most recent amendment, June 1989, Southampton: Ordnance Survey.

ORDNANCE SURVEY, 1990, *Ordnance Survey 1 : 250 000 Scale Data User Manual*, Southampton: Ordnance Survey.

ORDNANCE SURVEY, 1991, *Ordnance Survey 1 : 50 000 Scale Trial Data – Technical Specification*, Southampton: Ordnance Survey.

ORDNANCE SURVEY, 1992, *Ordnance Survey Digital Map Data Catalogue*, Southampton: Ordnance Survey.

PAINHO, M., 1995, The effects of generalization on attribute accuracy in natural resource maps, in MÜLLER, J.-C., LAGRANGE, J.-P. and WEIBEL, R. (Eds.), *GIS and Generalisation: Methodology and Practice*, London: Taylor & Francis, pp. 194–206.

PENCK, A., 1894, *Morphologie der Erdoberfläche*, Stuttgart (cited in Maling, 1989).

PERKAL, J., 1958, Proba obiektywnej generalizacji, *Geodezja i Kartografia*, 7(2), 130–42. ('An attempt at objective generalisation' – translated by W. Jackowski in 1965.)

PETCHENIK, B., 1983, A map maker's perspective on map design research 1950–1980, in TAYLOR, D. (Ed.), *Progress in Contemporary Cartography*, Vol. 2, New York: John Wiley, pp. 37–68.

PETRIE, G., 1990, Digital mapping technology: procedures and applications, in KENNIE, T. and PETRIE, G. (Eds.), *Engineering Surveying Technology*, Glasgow: Blackie, pp. 329–89.

PHILLIPS, R. and NOYES, L., 1982, An investigation of visual clutter in the topographic base of geological map, *The Cartographic Journal*, 19(2), 122–31.

PLAZANET, C., 1995, Measurements, characterization, and classification for automated line feature generalization, *ACSM/ASPRS Annual Convention and Exposition*, Charlotteville, North Carolina, Vol. 4 (Proceedings of AUTO-CARTO 12), pp. 59–68.

PLAZANET, C., AFFHOLDER, J.-G. and FRITSCH, E., 1995, The importance of geometric modeling in linear feature generalization, in WEIBEL, R. (Ed.), Special Issue, 'Map generalization', *Cartography and Geographic Information Systems*, 22(4), 291–305.

RAPER, J. and BUNDOCK, M., 1991, UGIX: a layer based model for a GIS user interface, in MARK, D. and FRANK, A. (Eds.), *Cognitive and Linguistic Aspects of Geographic Space*, Dordrecht: Kluwer Academic, pp. 449–75.

RAPER, J. and GREEN, N., 1989, Development of a hypertext based tutor for geographical information systems, *British Journal of Educational Technology*, **3**, 164–72.

RAPER, J., RHIND, D. and SHEPHERD, J., 1992, *Postcodes: the New Geography*, Harlow: Longman.

RHIND, D., 1973, Generalisation and realism within automated cartographic systems, *Canadian Cartographer*, **10**(1), 51–62.

RHIND, D., 1988, Personality as a factor in the development of a discipline: the example of computer-assisted cartography, *American Cartographer*, **15**(3), 277–90.

RHIND, D., 1990, Topographic data bases derived from small scale maps and the future of Ordnance Survey, in FOSTER, M. and SHAND, P. (Eds.), *The Association for Geographic Information Yearbook 1990*, London: Taylor & Francis, pp. 87–96.

RHIND, D. and CLARK, P., 1988, Cartographic data inputs to global databases, in MOUNSEY, H. and TOMLINSON, R. (Eds.), *Building Databases for Global Science*, London: Taylor & Francis.

RHIND, D., RAPER, J. and GREEN, N., 1989, First UNIX, then UGIX, *Proceedings of AUTO-CARTO 9*, Baltimore, Maryland, 2–7 April, pp. 735–44.

RICHARDSON, D., 1989, Rule based generalisation for map production, *Proceedings of the National Conference 'GIS – challenge for the 1990s'*, Ottawa, Canada, pp. 718–39.

RICHARDSON, L., 1961, The problem of contiguity, *General Systems Yearbook*, **6**, 139–87.

RIEGER, M. and COULSON, M., 1993, Consensus or confusion: cartographers' knowledge of generalization, *Cartographica*, **30**(1), 69–80.

ROBINSON, A., 1989, Cartography as an art, in RHIND, D. and TAYLOR, D. (Eds.), *Cartography Past, Present and Future*, London: Elsevier, pp. 91–102.

ROBINSON, A. and SALE, R., 1969, *Elements of Cartography*, 3rd edn, New York: John Wiley.

ROBINSON, A., SALE, R., MORRISON, J. and MUEHRCKE, D., 1984, *Elements of Cartography*, 5th edn, New York: John Wiley.

ROBINSON, G. and ZALTASH, A., 1989, Application of expert systems to topographic map generalisation, *Proceedings of AGI '89, 'GIS – A Corporate Resource'*, pp. A.3.1–6.

RUAS, A., 1995a, Multiple paradigms for automating map generalization: geometry, topology, hierarchical partitioning and local triangulation, *ACSM/ASPRS Annual Convention and Exposition*, Charlotteville, North Carolina, Vol. 4 (Proceedings of AUTO-CARTO 12), pp. 69–78.

RUAS, A., 1995b, Multiple representations and generalization, Lecture notes for Nordic Summer Course in Cartography (ftp://sturm.ign.fr).

RUAS, A. and PLAZANET, C., 1997, Strategies for automated generalization, in KRAAK, M.-J., MOLENAAR, M. and FENDEL, E. (Eds.), *Advances in GIS Research II* (Proceedings of the 7th International Symposium on Spatial Data Handling), London: Taylor & Francis, pp. 319–36.

SALGÉ, F., 1993, Is the rest of Europe any different? *Proceedings of the AGI Conference*, Birmingham, 16–18 November, pp. 1–4.

SCHLEGEL, A. and WEIBEL, R., 1995, Extending a general-purpose GIS for computer-assisted generalization, *Proceedings of the 17th International Car-*

tographic Conference: 'Cartography Crossing Borders', Barcelona: Institut Cartogrà- fic de Catalunya, pp. 2211–20.

SCHYLBERG, L., 1993, *Computational methods for generalization of cartographic data in a raster environment*, PhD thesis, Department of Geodesy and Photogrammetry, Royal Institute of Technology, Sweden, Photogrammetric Reports No. 60, Stockholm: Royal Institute of Technology.

SHEA, K. and MCMASTER, R., 1989, Cartographic generalization in a digital environment: when and how to generalize, *Proceedings of AUTO-CARTO 9*, Baltimore, Maryland, 2–7 April, pp. 56–67.

SHELBERG, M., MOELLERING, H. and LAM, N., 1982, Measuring the fractal dimensions of empirical cartographic curves, *Proceedings of AUTO-CARTO 5*, pp. 481–90.

SMALLWORLD SYSTEMS, 1994, *Smallworld GIS 2 User Guide*, Cambridge, UK: Smallworld Systems, Ltd.

SMITH, A., 1989, Small scale data – a large scale problem? *Proceedings of AGI '89, 'GIS – A Corporate Resource'*, pp. B.4.1–4.

SMITH, N., 1980, The automated generalisation of small scale topographic maps, with particular reference to the Ordnance Survey 1 : 50 000 scale series, Unpublished MA thesis, University of Durham.

SMITH, N. and RHIND, D., 1993, Defining the quality of spatial data: a discussion document, *Proceedings of the Land Information Management and GIS Conference*, University of New South Wales, July.

SOWTON, M., 1991, Development of GIS-related activities at the Ordnance Survey, in MAGUIRE, D., GOODCHILD, M. and RHIND, D. (Eds.), *Geographical Information Systems: Principles and Applications*, Vol. 2, Harlow: Longman, pp. 23–38.

STEINHAUS, H., 1954, Length, shape and area, *Colloquium Mathematicum*, **3**, 1–13.

SWISS SOCIETY OF CARTOGRAPHY, 1987, *Cartographic Generalisation – Topographic Maps*, Cartographic Publication Series, no. 2, 2nd edn, Zürich: Swiss Society of Cartography.

THAPA, K., 1988, Automatic line generalisation using zero-crossings, *Photogrammetric Engineering and Remote Sensing*, **54**(4), 511–17.

THAPA, K. and BOSSLER, J., 1992, Accuracy of spatial data used in geographic information systems, *Photogrammetric Engineering and Remote Sensing*, **58**(6), 835–41.

TIMES BOOKS, 1990, *The Times Concise Atlas of the World*, London: Times Books.

TOBLER, W., 1989, Frame independent spatial analysis, in GOODCHILD, M. and GOPAL, S. (Eds.), *Accuracy of Spatial Databases*, London: Taylor & Francis, pp. 115–22.

TÖPFER, F. and PILLEWIZER, W., 1966, The principles of selection, *Cartographic Journal*, **3**(1), 10–16.

UNWIN, D., 1981, *Introductory Spatial Analysis*, London: Methuen.

VAN OOSTEROM, P., 1993, *Reactive Data Structures for Geographic Information Systems*, Oxford: Oxford University Press.

VAN OOSTEROM, P. and SCHENKELAARS, V., 1995, The development of a multi-scale GIS, *International Journal of Geographical Information Systems*, **9**(5), 489–508.

VAUGHAN, D., 1991, Project-based environmental GIS, *Proceedings of AGI '91*, Birmingham, pp. 2.21.1–5.

VISVALINGAM, M. and WHYATT, J., 1990, The Douglas-Peucker algorithm for line simplification: re-evaluation through visualization, *Computer Graphics Forum*, **9**(3), 213–28.

VISVALINGAM, M. and WHYATT, J., 1993, Line generalisation by repeated elimination of points, *Cartographic Journal*, **30**(1), 46–51.

VITEK, J., WALSH, S. and GREGORY, M., 1984, Accuracy in GIS: an assessment of inherent and operational errors, *PECORA IX Symposium*, October.

VOROB'EV, V., 1959, Dlina beregovoy linii morey SSSR, *Geograficheskii Sbornik*, **XIII**, 63–89. ('The length of the coastline of the USSR' – cited in Maling, 1989.)

WALKER, P. (ed.), 1988, *Chambers Science and Technology Dictionary*, Edinburgh and Cambridge: W & R Chambers and Cambridge University Press.

WEHDE, M., 1982, Grid cell size in relation to errors in maps and inventories produced by computerized processing, *Photogrammetric Engineering and Remote Sensing*, **48**(8), 1289–98.

WEIBEL, R., 1986, Automated cartographic generalisation, in SIEBER, R. and BRASSEL, K. (Eds.), *A Selected Bibliography on Spatial Data Handling: Data Structures, Generalisation, and Three-dimensional Mapping*, Vol. 6, Geo-Processing Series, Department of Geography, University of Zürich, pp. 20–35.

WEIBEL, R., 1989, Design and implementation of a strategy for adaptive computer-assisted terrain generalization, *14th ICA World Conference*, 17–24 August, Budapest.

WEIBEL, R., 1991, Amplified intelligence and rule-based systems, in BUTTENFIELD, B. and McMASTER, R. (Eds.), *Map Generalization: Making Rules for Knowledge Representation*, Harlow: Longman, pp. 172–86.

WEIBEL, R., 1992, Models and experiments for adaptive computer-assisted terrain generalisation, *Cartography and Geographic Information Systems*, **19**(3), 133–53.

WEIBEL, R., 1995a, Map generalization in the context of digital systems, in WEIBEL, R. (Ed.), Special Issue, 'Map generalization', *Cartography and Geographic Information Systems*, **22**(4), 259–63.

WEIBEL, R., 1995b, Three essential building blocks for automated generalisation, in MÜLLER, J.-C., LAGRANGE, J.-P. and WEIBEL, R. (Eds.), *GIS and Generalisation: Methodology and Practice*, London: Taylor & Francis, pp. 56–69.

WEIBEL, R. and BUTTENFIELD, B., 1988, Map design for geographic information systems, *Proceedings of GIS/LIS '88*, San Antonio, Texas, 30 November–2 December, Vol. 1, pp. 350–9.

WEIBEL, R. and EHRLIHOLZER, R., 1995, An evaluation of MGE Map Generalizer: interim report, Department of Geography, University of Zürich.

WEIBEL, R., KELLER, S. and REICHENBACHER, T., 1995, Overcoming the knowledge acquisition bottleneck in map generalization: the role of interactive systems and computational intelligence, in FRANK, A. and KUHN, W. (Eds.), *Proceedings of COSIT '95: 'Spatial Information Theory – A Theoretical Basis for GIS'*, Lecture Notes in Computer Science 988. Berlin: Springer-Verlag, pp. 139–56.

WHITE, E., 1985, Assessment of line generalization algorithms using characteristic points, *The American Cartographer*, **12**(1), 17–28.

WOODSFORD, P., 1995, Object-orientation, cartographic generalisation and multi-product databases, *Proceedings of the 17th International Cartographic Conference: 'Cartography Crossing Borders'*, Barcelona: Institut Cartogràfic de Catalunya, pp. 1054–8.

XIA, Z.-G., CLARKE, K. and HUANG, J., 1991a, An evaluation of algorithms for estimating the fractal dimension of topographic surfaces, *ACSM–ASPRS Fall Convention 1991 – Technical Papers*, Atlanta, Georgia, 28 October–1 November, pp. A-164–80.

XIA, Z.-G., CLARKE, K. and PLEWS, R., 1991b, The uses and limitations of fractal geometry in terrain modelling, *ACSM–ASPRS'91*, Atlanta, Georgia, 28 October–1 November, pp. 336–51.

Author Index

Subject Index